Gold Nanoparticles - Reaching New Heights

Edited by Mohammed Rahman and Abdullah Mohammed Asiri

Published in London, United Kingdom

IntechOpen

Supporting open minds since 2005

Gold Nanoparticles - Reaching New Heights
http://dx.doi.org/10.5772/intechopen.73366
Edited by Mohammed Rahman and Abdullah Mohammed Asiri

Contributors
Ziyauddin Qureshi, Jaseer E.A., Sivakumar Thiripuranthagan, Pablo Gavina, Ana M. Costero, Margarita Parra, Salvador Gil, Samuel Briggs, Khalid Hattar, Young-Seok Shon, Ting-An Chen, Julián Puszkiel, Mohammed Muzibur Rahman

First published in London, United Kingdom, 2019 by IntechOpen
IntechOpen is the global imprint of INTECHOPEN LIMITED, registered in England and Wales, registration number: 11086078, The Shard, 25th floor, 32 London Bridge Street
London, SE19SG - United Kingdom
Printed in Croatia

British Library Cataloguing-in-Publication Data
A catalogue record for this book is available from the British Library

Additional hard copies can be obtained from orders@intechopen.com

Gold Nanoparticles - Reaching New Heights
Edited by Mohammed Rahman and Abdullah Mohammed Asiri
p. cm.
Print ISBN 978-1-78984-998-1
Online ISBN 978-1-78984-957-8

For EU product safety concerns:
IN TECH d.o.o., Prolaz Marije Krucifikse Kozulić 3, 51000 Rijeka, Croatia, info@intechopen.com or visit our website at intechopen.com.

Meet the editors

Mohammed Muzibur Rahman received his BSc and MSc from Shahjalal University of Science and Technology, Sylhet, Bangladesh, in 1999 and 2001, respectively. He received his PhD from the Chonbuk National University, South Korea, in 2007. After his PhD, he worked as postdoctoral fellowship and assistant professor in pioneering research centers and universities located in South Korea, Japan, and Saudi Arabia (2007–2011). Since 2011 he has been working as associate professor in the Center of Excellence for Advanced Materials Research and the Chemistry Department at King Abdulaziz University, Saudi Arabia. He has published more than 320 international and domestic conferences, several book chapters, and 10 books as an editor. His research work has been largely in the area of nanoparticles, carbon nanotubes, nanotechnology, sensors, ionic liquid, surface chemistry, electrochemistry, instrumental science, nanomaterials, self-assembled monolayers, photochemistry, m-chips and devices, etc.

Abdullah Mohamed Asiri received his PhD from the University of Wales, College of Cardiff, UK, in 1995. He has been the Head of the Chemistry Department at King Abdul Aziz University since October 2009 and is the founder and Director of the Center of Excellence for Advanced Materials Research. He is a professor of organic photochemistry. His research interest covers color chemistry, synthesis of novel photochromic and thermochromic systems, synthesis of novel coloring matters and dyeing of textiles, materials chemistry, nanochemistry, nanotechnology, polymers, and plastics. He is the Editor-in-Chief of King Abdul Aziz University Journal of Science. He is also a member of the Editorial Board of Pigments & Resin Technology (UK), Organic Chemistry Insights (New Zealand), and Recent Patents on Materials Science (USA). He is Vice-President of the Saudi Chemical Society (Western Province Branch).

Contents

Preface

It gives us immense pleasure to introduce this book called *Gold Nanoparticles—Reaching New Heights*, based on the state of the art of gold nanoparticles (AuNPs) with their outstanding and potential applications. The book deals with the advanced nanotechnological aspects of synthesis, characterization, development, and potential optical and biological applications of AuNP materials. Discussion of these aspects develops through the fundamental and applied experimental routes using conventional methods via the interaction of AuNPs and finally brings together both scientific and technological worlds. Basically, AuNPs have undoubtedly achieved many accomplishments in a conventional sense and have taken new directions from preparation to practical applications in research and development in different areas of science and technology. New paths and emerging frontiers branch out from time to time from this advanced nanotechnology stage of low nanodimensional AuNPs. Advances in AuNPs with instrumentation for evaluating the structural model in aqueous or non-aqueous phases now enable us to understand quite broadly almost all the events that take place with gold AuNPs at least at the nano level.

In this book, authors Gavino et al. focus on the design of colorimetric sensors and probes due to their interesting photophysical properties. In this approach, the surface plasmon resonance (SPR) band is sensitive to the proximity of other nanoparticles and thus analyte-triggered aggregation of AuNPs results in an important bathochromic shift of the SPR band and a change in the color of the solution from red to blue due to interparticle surface plasmon coupling. The selectivity of AuNP-based sensors towards different analytes depends on the recognition properties of the molecules attached to the surface of the nanoparticles. Finally, a selection of biologically active molecules is considered as analytes: neurotransmitters, nerve agents, pesticides, and carboxylates of biological interest. Taking into account the interesting photophysical properties of AuNPs, their easy functionalization, the use of aqueous solutions, and detection using the naked eye, they conclude that the red or blue question will continue to be ever present in the molecular sensing field.

Qureshi et al. also approach the higher catalytic activity being observed for AuNPs supported on reducible metal oxides such as TiO_2, CO_3O_4, CeO_2, and Fe_2O_3. Here, they study in detail CO oxidation catalyzed by mono- and bimetallic AuNPs over various silica supports. Finally, it is true to say that silica-supported gold nanocatalysts are becoming a hot topic of research for CO oxidation; conversely, there are still many challenges ahead for the improvement of silica-supported gold nanocatalysts to fulfill the main necessities of any catalyst such as an easy and low-cost synthesis method, high activity, selectivity, and greater stability at lower temperatures. Furthermore, they conclud that the proof of identity for active gold species is still a challenging task for CO oxidation reactions catalyzed by gold.

Briggs et al. approach nanoparticle interactions with energetic neutrons, photons, and charged particles that can cause structural damage ranging from single atom displacement events to bulk morphological changes. Due to the diminutive length scales and prodigious surface-to-volume ratios of AuNPs, radiation damage effects are typically dominated by sputtering and surface interactions and can vary drastically from bulk behavior and classical models. Here, they report on contemporary experimental and computational modeling efforts that have contributed to the current understanding of how ionizing radiation environments affect the structure and properties of AuNPs. Finally, the future potential for elucidating the active mechanisms in AuNPs exposed to ionizing radiation and the subsequent ability to predictively model the radiation stability and ion beam modification parameters are discussed in his chapter.

Sivakumar et al. develop a promising technique for photodegradation of various hazardous chemicals that are encountered in waste waters. Here, they investigate the performance of various semiconductors. The anatase phase of TiO_2 affords the best compromise between catalytic performance and stability in aqueous medium. Apart from many positive attributes of TiO_2, the main drawbacks associated with this catalyst are (1) large band gap ($E_g > 3.2$ eV), which can be excited only by UV light, and (2) recombination of excitons. Dye sensitization, coupling of semiconductors, transitional metal doping, etc., are some methods to shift the optical response from UV to the visible spectral range. They also introduce another important approach to shift the optical response of TiO_2 from UV to the visible spectral range, i.e., by doping of noble metals with TiO_2. Here, they research nanoparticles of different noble metals such as Ag, Au, and Pt deposited on synthesized TiO_2, characterized by using various instrumental techniques such as XRD, TEM, FT-IR, BET, UV-Vis, and AAS and subjected to the degradation of textile dyes, namely TAZ, RY-17, and RB-5 under both UV and visible irradiations. Reaction conditions such as catalyst concentration, dye concentration, pH, irradiation time, light intensity, and additives were optimized for complete decolorization and are discussed in detail in this chapter.

Chen and Shon review intriguing catalytic studies that are accomplished by employing a variety of catalysts such as metal complexes, supported materials, supported metal complexes, and nanosized materials for polyene hydrogenation. Additionally, unsupported colloidal nanoparticle catalysts, which exhibit excellent activity and selectivity toward polyene hydrogenation, are introduced. The high activity of colloidal metal nanoparticle catalysts often allows reactions to be completed under mild conditions at atmospheric pressure and at room temperature. The important fundamental understandings of the influence of chemical environments (solvents, ligands, dopants, etc.) and compositions (metal complexes, metals, alloys, etc.) towards the catalytic activity and selectivity of various catalysts in homogeneous, heterogeneous, and semi-heterogeneous conditions are discussed. Systematic evaluation is also discussed in this review chapter, which paves the way to further develop chemo-, regio-, and stereoselective catalysts for polyene hydrogenation.

Finally, Puszkiel discusses hydride-forming binary as well as complex materials for their potential hydrogen storage properties, which possess high volumetric and gravimetric hydrogen capacities. Tuning the kinetic behavior of these hydride-forming materials involves different approaches and their combinations are discussed in this chapter. Herein, basic concepts of the chemical reaction for hydride compound formation/decomposition, thermodynamics, kinetics, and applied strategies to enhance the kinetic behavior of hydride compounds and systems are comprehensively described and discussed.

This work aims to bridge the gap between undergraduates, graduates, and scientists in applied AuNPs as well as modified colloidal gold particles to initiate researchers to study in as straightforward a way as possible and to introduce them to the opportunities offered by applied science and technological fields. We worked unswervingly to complete this work with the help of IntechOpen Open Access publisher. We hope that this contribution will further enhance applied AuNPs in nano- and bioscience, especially in bringing new entrants into the applied and hybrid AuNP science and technology fields and help scientists to forward and develop their own fields of specialization.

Mohammed Rahman and Abdullah Mohammed Asiri
King Abdulaziz University,
Kingdom of Saudi Arabia

Section 1

State-of-Art Gold Nanoparticles

Chapter 1

Introductory Chapter: Basic Concept of Gold Nanoparticles

Mohammed Muzibur Rahman and Abdullah Mohamed Asiri

1. Introduction

Nanoscience and nanotechnology are generally used in the design, production, characterization, and potential applications of nanostructural materials especially considering their size as well as shape. Nanoscience is a phenomenon that occurs in structures of nanodimensions. Generally, the unique features of nanosystems arise exclusively from the small size of the systems. Here, nano is the smallest dimension as it is obtained in the world in various branches of chemistry, physics, drug design, semiconductor materials science, and even biological science. The hydrogen atom diameter is about 1/10 of a nanometer; therefore, the nanometer dimension or scale is the very tiny scale on what we might consider the building objects or machines on the fundamental basis of the principles in where from everyday mechanics. By using the 1000 as well as hydrogen atoms it could be picked into a cubic object. Nanoscience and nanotechnology are cumulative designations referring to each practical technology and instrumental science, which function with nanodimensional scale or objects. Basically, low-dimensional nanoparticles have various significant properties compared to those of larger objects/particles, and these characteristic properties could be utilized in a broad spectrum of areas of medicine, catalysis, information technologies, renewable energy production, renewable energy storage, ultrasensitive sensors, devices, materials, manufacturing, surfactants, and environmental applications. Basically, the development of green nanotechnology is generating interest in researchers toward eco-friendly, safe, and non-toxic routes of synthesis that can be used for manufacturing at a large scale. This is a simple, cost-effective, stable for long time, and reproducible aqueous room temperature synthesis method to obtain a self-assembly of gold nanoparticles.

2. Literature survey

Generally, colloidal gold is a kind of sol or colloidal suspension, which is called nanoparticles of gold in a water fluid. Colloidal gold nanoparticles usually show either an intense red or blue/purple color. Molecular sensors are generally composed of two main elements: a recognition unit, which selectively interacts with the molecule or ion to be detected, and a signaling unit that is responsible of generating a macroscopically measurable signal (optical, electrochemical, or mechanical) upon the molecular recognition event. In general, the term chemosensor is used when the binding between the receptor and the target analyte is reversible and probe, when the recognition event takes place through an irreversible chemical reaction. The transduction mechanism, which is the mechanism by which the chemical interaction between the analyte and the recognition unit is converted into a change in the macroscopic signal of the signaling unit, will depend on the type of signaling unit and the structure of the receptor. The use of chemosensors for the selective colorimetric detection of small biologically active molecules offers some advantages over traditional instrumental analytical methods. Chromogenic probes are usually cheap and easy to use, and they do not require expensive instrumentation; very often, the presence of the analyte can be detected by the naked eye, which allows for rapid in-situ detection. Among the different approaches for colorimetric sensing, the use of gold nanoparticles (AuNPs) as scaffolds and signaling units for the construction of molecular sensors has attracted enormous interest for several reasons: they can be easily synthesized from Au(III) salts in various sizes and shapes. Their surface can be functionalized with a wide range of thiol- or disulfide-terminated organic ligands, by ligand exchange reactions, through the formation of strong Au—S bonds, leading to stable colloidal suspensions in water or organic solvents (depending on the ligand). Finally, gold nanoparticles have remarkable optoelectronic properties. In particular, the localized surface plasmon resonance (SPR) is influenced by not only the size and shape of nanoparticles but also by the dielectric properties of the environment and the proximity of other nanoparticles [1, 2]. Among the neurotransmitters, biogenic selective amines are paid specific attention due to their influence in areas ranging from biomarkers of specific problems or diseases [3–5] to quality control of foodstuffs [6, 7]. Nitric oxide (NO) is also a very important neurotransmitter in the central, peripheral, and enteric nervous systems (ENS) [8, 9]. Dopamine, the simplest biogenic catecholamine (CA), is an important neurotransmitter of the central and peripheral nervous systems [10]. An approach for the colorimetric detection of dopamine has been developed using 4-amino-3-hydrazino-5-mercapto-1,2,4-triazole functionalized AuNPs. Dopamine induced the aggregation of the AuNPs through hydrogen bonding interactions [11]. Each dopamine molecule has three H-donor groups able to form hydrogen bonds (the amino and both hydroxyl groups). The recent rise in international apprehension over the use of chemical warfare (CW) agents in different conflictive scenarios has resulted in an increasing attention in the determination of these lethal chemicals. Among CW species, nerve agents are greatly treacherous, and their high toxicity and easy construction underscore the need to detect these deadly chemicals via quick and reliable procedures. AuNPs have been used as sensors for some nerve agent simulants with good results. Different sensing mechanisms have been used for detecting these compounds; for example, compensation of charges has been applied for this process [12]. The design sensing protocol takes advantage of the nucleophilic reactivity of pyridine moieties toward nerve gases [13].

Here, scattering interactions are more common for higher energy neutron spectra. Due to the large difference in mass between neutrons and gold nuclei, neutrons undergoing elastic scattering interactions (i.e., ballistic collisions) only transfer approximately 1% of their energy to the nucleus [14]. Inelastic scattering interactions, though less common, can transfer significantly more energy, leaving the nucleus in an excited state that results in gamma ray production. With sufficiently energetic incident neutrons, both scattering mechanisms are capable of generating gold primary knock-on atoms and generating displacement damage in the material microstructure. Radiation therapy is likely one of the most notable examples in which gold nanoparticle injections in the vicinity of a tumor or conjugated to antibodies such that they preferentially bind to cancer cells have been shown to increase the local radiation dose during X-ray irradiation through secondary photon interactions, thus lessening the absorbed dose in the surrounding healthy tissue [15–17]. A similar local dose enhancement effect has been shown for proton and heavy-ion irradiation therapies, which are being pursued to further reduce the dose to healthy tissue as a consequence of how energetic charged particles deposit a majority of their energy at the end of their range [18, 19]. Furthermore, potential applications in satellites or space electronics would involve bombardment with high-energy cosmic radiation [20, 21].

On the other hand, ionizing radiation is known to have the ability to drastically alter the microstructure and performance of materials primarily through the displacement of constituent atoms from their lattice sites, resulting in the generation of damage and point defects [22]. These point defects tend to diffuse and coalesce into larger, ordered defect structures such as dislocation loops, cavities, and stacking faults, typically resulting in deleterious effects in structural materials such as radiation-induced hardening and embrittlement or void swelling. Irradiation can also cause solute redistribution in alloys and composite materials, encouraging the precipitation of secondary phases or promoting localized corrosion, which has been utilized in metal, ceramics, and polymers to provide added functionality not previously possible [23–25]. The fundamental mechanisms of radiation damage and the subsequent effects on the properties and performance of materials have been and continue to be heavily studied in commonly used and candidate materials for nuclear reactors and extraterrestrial applications. The mechanism of radiation damage in materials with limited dimensions has been shown to be significantly different from bulk material behavior [26, 27]. Due to the prodigious surface-to-volume ratio of nanoparticles compared to even thinner films, they have emerged as an attractive material choice for many applications. However, this same surface-to-volume ratio results in sputtering and free surface effects ultimately dominating the radiation response [28, 29]. Put simply, instead of diffusing into organized defect structures, the generated point defects tend to annihilate at the particle surface, or the material is ejected from the particle volume, as a result of the energetic collision [30, 31]. This can drastically alter the shape of individual particles and cause agglomeration of closely-spaced particle groupings [32–34]. Such morphological changes have the potential to adversely affect their efficacy in applications where the high surface-to-volume ratio and local structure are essential for performance.

The three primary interactions of photons with matter include the photoelectric effect, Compton scattering, and pair production [35]. In a photoelectric event, a photon completely transfers its energy to an orbital electron, ejecting it from its shell and ionizing the atom. This is the dominant interaction mechanism for low-energy photons. In Compton scattering, an incident photon transfers a portion of its energy to an orbital electron, and both the electron and photon continue typically with different trajectories when compared to the incoming photon. This will be the dominant interaction mechanism for incident photons of intermediate energies. Finally, pair production dominates for high energy photons and can only occur for incident photons greater than 1.022 MeV. In pair production, a photon interacts with the electromagnetic field surrounding the atomic nucleus and transforms into an electron and a positron, with the total photon energy less than the remaining energy of the two particles (0.511 MeV each) being shared between them. It should be noted that for high photon energies (>~8 MeV), photodisintegration can also occur, in which the incident photon causes the atom to emit one or more neutrons, potentially resulting in a radioactive isotope of gold [36]. Additionally, photon irradiation experiments involving gold nanoparticles are prevalent, as X-ray therapies for cancer treatment are fairly common practice [37–39]. While these studies are predominantly focused on the biological effects in the vicinity of cancerous tissue, they highlight how photons interact with these high-atomic-number (high-Z) nanoparticles to deposit their energy locally. High energy photons are also expected to cause Frenkel pair displacement damage (i.e., vacancy and interstitial defect pairs), primarily through the energetic electrons that they tend to generate. However, the effects of these displacement effects on the long-term stability of nanoparticle structures, if any, have not been studied to our knowledge. Charged particles are common by-products of radioactive decay and nuclear reactions and are a primary component of cosmic radiation. Accelerators and ion beams are also common ion sources used in both research and industry. For example, ions are frequently used in radiation damage experiments to simulate material microstructures resulting from neutron radiation exposure in nuclear reactor environments primarily because they are much less costly and can achieve similar damaged microstructures in a fraction of time [40]. Accelerators also have a slew of other potential applications ranging from materials analysis (e.g., electron microscopes and Rutherford backscattering) to ion beam modification. Beta (electron or positron) radiation either incident on or produced in gold will primarily lose its energy via ionization and bremsstrahlung radiation. As mentioned in the introduction, bremsstrahlung radiation occurs due to electron acceleration from interaction with an atomic nucleus and results in the production of a photon of energy equivalent to the energy lost by the electron. This interaction is again Z-dependent and is quite common for high-Z materials like gold [41]. While displacement or knock-on damage resulting from electrons is commonly observed in transmission electron microscopy experiments [42], the significant difference in the masses of an electron and a gold nucleus requires electron energies in excess of 1.35 MeV to create a single Frenkel pair [43]. In many cases, the role of the electron beam in altering the nanoparticle stability during these studies is not a result of interaction with the gold atoms itself, but with the organic capping ligands, as a result, the stability of the gold nanoparticles to beta radiation is often dictated by the organic capping agent chosen [44]. Alpha particles are common products of radioactive decay for actinides and other heavy radioactive isotopes, but, along with protons, deuterons, and tritons, they can also result from and induce a number of different nuclear reactions. For example, irradiation of Au-197 with protons with energies of 4.5 MeV or higher can cause a (p, n) reaction resulting in the production of metastable Hg-197 m [45]. However, the threshold energies of these reactions for gold are, in most cases, sufficiently high and reaction cross sections sufficiently low that these types of interactions rarely occur in practical applications. More often, light ions will interact via ionization and through coulombic forces. Similar to the other types of interactions discussed, ionization has the potential to result in the emission of characteristic X-rays and other secondary radiation. Coulombic interactions with other atomic nuclei can tend to cause displacement damage, usually in the form of Frenkel pairs or small, isolated cascades.

In general, the response of nanostructured materials to radiation damage is still poorly understood [46]. Despite the limited understanding in the general field, free-standing gold nanoparticles (usually drop-cast onto carbon or SiN TEM grids) have been used as the model system for testing and validation of TEM with in-situ ion irradiation capabilities [47–49]. Expanding on the known enhanced sputtering rate observed in gold thin foils exposed to a range of noble gas ions [50], it was later shown by Ilinov et al. that gold nanorods irradiated with 80 keV Xe demonstrated sputtering rates three orders of magnitude higher than those predicted by classical sputtering simulations [33]. This surprising set of results means that classical models and expectations no longer hold true when predicting the radiation response of gold nanoparticles. The response of both free-standing and embedded gold nanoparticles is drastically different from that of gold in bulk or thin-film morphologies and is highly dependent on the radiation environment, particle morphology, and surface conditions. Additional work is needed to elucidate the underlying physics governing the increased sputtering and other scaling effects observed in gold nanoparticles [32, 33, 51]. Without a detailed understanding of the mechanisms active when the displacement damage length scale of the radiation event approaches that of the size of the nanoparticle exposed, it will be challenging to employ gold nanoparticles in most radiation environments. If the significant enhancement of sputtering is inherent and cannot be overcome, then the application of gold nanoparticles subject to ionizing radiation may be limited to those environments that produce minimal dose or sputtering yields.

3. Conclusions

Gold nanoparticles have diverse properties compared to those of bigger particles, which can be exploited in a broad spectrum of fields such as in renewable energy, nanomedicine, catalysis, information technologies, energy production, energy storage, sensitive nanosensors, nanomaterials, manufacturing, and real-time environmental applications. Taking into account the interesting photo-physical properties of sensitive gold nanoparticles, their easy functionalization, the use of aqueous solutions, and the detection by the naked eye, it can be concluded that the red or blue signal will continue to be very present in the molecular sensing field. It is concluded that selective gold nanoparticles have been extensively used for the design of colorimetric sensors as well as probes due to their interesting photo-physical properties. The use of gold nanoparticles for the preparation of colorimetric sensors is a very active field. The changes in the color of colloidal gold nanoparticles in solution because of the change in the surface plasmon absorption band upon aggregation or disaggregation processes can be easily used to transform the molecular recognition event into a macroscopic measurable signal. This change from red to blue can be perceived by the naked eye allowing in this way cheap and easy detection of the target analytes. The initial studies that have been done to observe the radiation response of gold nanoparticles to charged particle irradiation indicate that a significant enhancement of sputtering yield is present. This enhanced sputtering leads to the rapid disintegration of the original nanoparticle and the formation of unique satellite nanoscale arrangements as well as neurotransmitters, nerve agents, pesticides, and carboxylates of biological interest. The sensitivity and selectivity of the gold nanoparticle–based sensors toward the different bio- or chemical analytes will depend on the detection as well as recognition properties of the selective molecules attached to the surface of the low-dimensional nanoparticles. In this book, a selection of optically as well as biologically active molecules have been considered as analytes in broad interest such as photocatalysis, electrocatalysis, plasma-absorption, neurotransmitters, nerve agents, pesticides, and carboxylates.

Author details

Mohammed Muzibur Rahman* and Abdullah Mohamed Asiri
Chemistry Department, Faculty of Science, Center of Excellence for Advanced Materials Research (CEAMR), King Abdulaziz University, Jeddah, Saudi Arabia

*Address all correspondence to: mmrahmanh@gmail.com

References

[1] Saha K, Agasti SS, Kim C, Li X, Rotello VM. Gold nanoparticles in chemical and biological sensing. Chemical Reviews. 2012;**112**:2739-2779

[2] Mayer KM, Hafner JH. Localized surface plasmon resonance sensors. Chemical Reviews. 2011;**111**:3828-3857

[3] Marecos C, Ng J, Kurian MA. What is new for monoamine neurotransmitter disorders? Journal of Inherited Metabolic Disease. 2014;**37**:619-626

[4] Plonka J. Methods of biological fluids sample preparation—Biogenic amines, methylxanthines, water-soluble vitamins. Biomedical Chromatography. 2015;**29**(1):10

[5] Bachrach U. Polyamines and cancer: Minireview article. Amino Acids. 2004;**26**:307-309

[6] Jairath G, Singh PK, Dabur RS, Rani M, Chaudhari M. Biogenic amines in meat and meat products and its public health significance: A review. Journal of Food Science and Technology. 2015;**52**:6835-6846

[7] Guo YY, Yang YP, Peng Q, Hang Y. Biogenic amines in wine: A review. International Journal of Food Science and Technology. 2015;**50**:1523-1532

[8] Goyal RK, Hirano I. The enteric nervous system. The New England Journal of Medicine. 1996;**334**:1106-1115

[9] Stark ME, Szurszewski JH. Role of nitric oxide in gastrointestinal and hepatic function and disease. Gastroenterology. 1992;**103**:1928-1949

[10] Adams RN. Probing brain chemistry with electroanalytical techniques. Analytical Chemistry. 1976;**48**:1126A-1138A

[11] Feng JJ, Guo H, Li YF, Wang YH, Chen WY, Wang AJ. Single molecular functionalized gold nanoparticles for hydrogen-bonding recognition and colorimetric detection of dopamine with high sensitivity and selectivity. ACS Applied Materials and Interfaces. 2013;**5**:1226-1231

[12] Martí A, Costero AM, Gaviña P, Gil S, Parra M, Brotons-Gisbert M, et al. Functionalized gold nanoparticles as an approach to the direct colorimetric detection of DCNP nerve agent simulant. European Journal of Organic Chemistry. 2013;**2013**:4770-4779

[13] Royo S, Costero AM, Parra M, Gil S, Martinez-Mañez R, Sancenón F. Chromogenic, specific detection of the nerve-agent mimic DCNP (a Tabun Mimic). Chemistry—A European Journal. 2011;**17**:6931-6934

[14] Was GS. Fundamentals of Radiation Materials Science. 2nd ed. New York: Springer-Verlag; 2017

[15] Ferrari M. Cancer nanotechnology: Opportunities and challenges. Nature Reviews Cancer. 2005;**5**:161-171

[16] Kwatra D, Venugopal A, Anant S. Nanoparticles in radiation therapy: A summary of various approaches to enhance radiosensitization in cancer. Translational Cancer Research. 2013;**2**(4):330-342

[17] Stathakis S. The physics of radiation therapy. Medical Physics. 2010;**37**(3):1374-1375

[18] Kim JK et al. Therapeutic application of metallic nanoparticles combined with particle-induced x-ray emission effect. Nanotechnology. 2010;**21**(42):42510

[19] Lin Y et al. Biological modeling of gold nanoparticle enhanced

radiotherapy for proton therapy. Physics in Medicine and Biology. 2015;**60**(10):4149-4168

[20] Holmes-Siedle A, Adams L. Handbook of Radiation Effects. UK: Oxford University Press; 1993

[21] Ma TP, Dressendorfer PV. Ionizing Radiation Effects in MOS Devices and Circuits. USA: Wiley-Interscience; 1989

[22] Rodriguez P, Krishnan R, Sundaram CV. Radiation effects in nuclear reactor materials—Correlation with structure. Bulletin of Materials Science. 1984;**6**(2):339-367

[23] Busby JT, Was GS, Kenik EA. Isolating the effect of radiation-induced segregation in irradiation-assisted stress corrosion cracking of austenitic stainless steels. Journal of Nuclear Materials. 2002;**302**:20-40

[24] Jiao Z, Was GS. Novel features of radiation-induced segregation and radiation-induced precipitation in austenitic stainless steels. Acta Materialia. 2011;**59**(3):1220-1238

[25] Dearnaley G. Ion beam modification of metals. Nuclear Instruments and Methods in Physics Research Section B: Beam Interactions with Materials and Atoms. 1990;**50**(1-4):358-367

[26] Järvi TT et al. Damage production in nanoparticles under light ion irradiation. Physical Review B. 2009;**80**(13):1-4

[27] Järvi TT, Nordlund K. Sputtering of freestanding metal nanocrystals. Nuclear Instruments and Methods in Physics Research Section B: Beam Interactions with Materials and Atoms. 2012;**272**:66-69

[28] Krasheninnikov AV, Nordlund K. Ion and electron irradiation-induced effects in nanostructured materials. Journal of Applied Physics. 2010;**107**:1-70

[29] Li J et al. In situ heavy ion irradiation studies of nanopore shrinkage and enhanced radiation tolerance of nanoporous Au. Scientific Reports. 2017;7:39484

[30] Hoppe SM et al. Application of in-situ ion irradiation TEM and 4D tomography to advanced scintillator mateirals. SPIE. 2012;**8**:1-10

[31] Donnelly S, Birtcher R. Heavy ion cratering of gold. Physical Review B. 1997;**56**(21):13599

[32] Hinks JA et al. Effects of crystallographic and geometric orientation on ion beam sputtering of gold nanorods. Scientific Reports. 2018;**8**(1):512

[33] Ilinov A et al. Sputtering yields exceeding 1000 by 80keV Xe irradiation of Au nanorods. Nuclear Instruments and Methods in Physics Research Section B: Beam Interactions with Materials and Atoms. 2014;**341**:17-21

[34] Bufford DC et al. In situ TEM ion irradiation and implantation effects on Au nanoparticle morphologies. Chemical Communications (Cambridge, England). 2014;**50**(57):7593-7596

[35] Martin JE. Physics for Radiation Protection. Weinheim, Germany: Wiley-VCH; 2006

[36] Ermakov AN et al. Multineutron photodisintegration of the Au-197 nucleus behind the giant dipole resonance. Physics of Atomic Nuclei. 2008;**71**:397

[37] Hainfeld JF, Slatkin DN, Smilowitz HM. The use of gold nanoparticles to enhance radiotherapy in mice. Physics in Medicine and Biology. 2004;**49**(18):N309-N315

[38] Mesbahi A. A review on gold nanoparticles radiosensitization effect in radiation therapy of cancer. Reports

of Practical Oncology and Radiotherapy. 2010;**15**(6):176-180

[39] Shrestha S et al. Gold nanoparticles for radiation enhancement in vivo. Jacobs Journal of Radiation Oncology. 2016;**3**(1):026

[40] Was GS et al. Emulation of reactor irradiation damage using ion beams. Scripta Materialia. 2014;**88**:33-36

[41] Hippler R et al. Z dependence of Bremsstrahlung radiation from free atoms. Physical Review Letters. 1981;**46**(25):1622-1625

[42] Williams DB, Carter CB. Transmission Electron Microscopy: A Textbook for Materials Science. New York: Springer; 2009

[43] Bauer W, Sosin A. Point defect studies in gold by electron irradiation at low temperatures. I. Threshold displacement energy and displacement cross section. Physical Review. 1964;**135**(2A):A521-A526

[44] Egerton R, Li P, Malac M. Radiation damage in the TEM and SEM. Micron. 2004;**35**(6):399-409

[45] Elmaghraby EK et al. Production of the mercury-197 through proton induced reaction on gold. Applied Radiation and Isotopes. 2010;**68**(9):1694-1698

[46] Zhang X et al. Radiation damage in nanostructured materials. Progress in Materials Science. 2018;**96**:217-321

[47] Hinks J. A review of transmission electron microscopes with in situ ion irradiation. Nuclear Instruments and Methods in Physics Research Section B: Beam Interactions with Materials and Atoms. 2009;**267**(23-24):3652-3662

[48] Hinks JA. Transmission electron microscopy with in situ ion irradiation. Journal of Materials Research. 2015;**30**(9):1214-1221

[49] Hattar K, Bufford DC, Buller DL. Concurrent in situ ion irradiation transmission electron microscope. Nuclear Instruments and Methods in Physics Research Section B: Beam Interactions with Materials and Atoms. 2014;**338**:56-65

[50] Birtcher RC, Donnelly SE, Schlutig S. Nanoparticle ejection from gold during ion irradiation. Nuclear Instruments and Methods in Physics Research Section B: Beam Interactions with Materials and Atoms. 2004;**215**(1-2):69-75

[51] Greaves G et al. Enhanced sputtering yields from single-ion impacts on gold nanorods. Physical Review Letters. 2013;**111**(6):065504

Chapter 2

Red or Blue? Gold Nanoparticles in Colorimetric Sensing

Pablo Gaviña, Margarita Parra, Salvador Gil and Ana M. Costero

Abstract

Gold nanoparticles (AuNPs) have been extensively used for the design of colorimetric sensors and probes due to their interesting photophysical properties. In particular, their surface plasmon resonance (SPR) is influenced not only by the size but also by the shape or the properties of the matrix surrounding the nanoparticles. This SPR band is sensitive to the proximity of other nanoparticles, and thus, analyte-triggered aggregation of AuNPs results in an important bathochromic shift of the SPR band and a change in the color of the solution from red to blue due to interparticle surface plasmon coupling. The selectivity of the AuNPs-based sensors toward the different analytes will depend on the recognition properties of the molecules attached to the surface of the nanoparticles. In this chapter, a selection of biologically active molecules has been considered as analytes: neurotransmitters, nerve agents, pesticides, and carboxylates of biological interest.

Keywords: gold nanoparticles, aggregation, neurotransmitter, nerve agents, pesticides, carboxylates

1. Introduction

A molecular sensor (or chemosensor) is a molecule or molecular ensemble designed to indicate the presence of a specific analyte through a detectable change in a macroscopic measurable signal. Molecular sensors are generally composed of two main elements: a recognition unit, which selectively interacts with the molecule or ion to be detected, and a signaling unit that is responsible of generating a macroscopically measurable signal (optical, electrochemical, mechanical...) upon the molecular recognition event. In general, in a chemosensor, the receptor binds the target analyte in a reversible manner. If the recognition event takes place through an irreversible chemical reaction, the chemosensor is better described as "probe." The transduction mechanism, which is the mechanism by which the chemical interaction between the analyte and the recognition unit is converted into a change in the macroscopic signal of the signaling unit, will depend on the type of signaling unit and the structure of the receptor.

The use of chemosensors or probes for the colorimetric detection of small biologically active molecules offers some advantages over traditional instrumental analytical methods. Chromogenic probes are usually cheap, easy to use, and do not require of expensive instrumentation, and very often, the presence of the analyte can be detected by the naked eye, which allows for rapid *in-situ* detection.

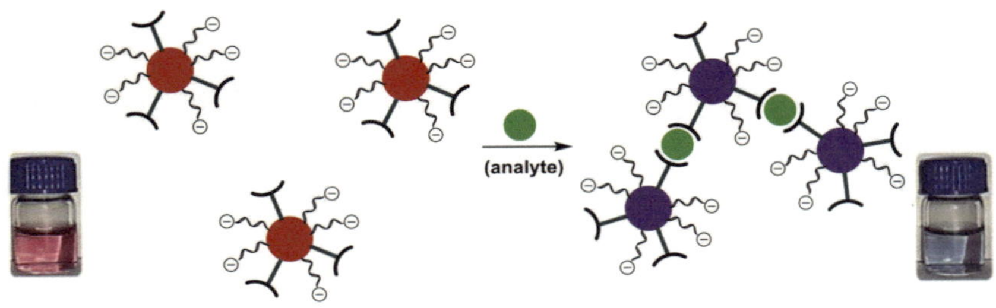

Figure 1.
Analyte-induced aggregation of gold nanoparticles with concomitant changes in the color of the colloidal suspensions.

Among the different approaches for colorimetric sensing, the use of gold nanoparticles (AuNPs) as scaffolds and signaling units for the construction of molecular sensors has attracted enormous interest for several reasons: They can be easily synthesized from Au(III) salts in various sizes and shapes. Their surface can be functionalized with a wide range of thiol- or disulfide-terminated organic ligands, by ligand exchange reactions, through the formation of strong Au–S bonds, leading to stable colloidal suspensions in water or organic solvents (depending on the ligand). Finally, gold nanoparticles have remarkable optoelectronic properties. In particular, the localized surface plasmon resonance (SPR) gives rise to a strong absorption band in the visible region. This SPR band is influenced not only by the size and shape of the nanoparticles but also by the dielectric properties of the environment and the proximity of other nanoparticles [1, 2]. This last property is the basis for colorimetric assays using AuNPs. Thus, analyte-triggered aggregation of AuNPs results in an important bathochromic shift of the SPR band (from ca. 520 to ca. 650 nm) and a change in the color of the colloidal solution from red (dispersed) to blue (aggregated) due to interparticle surface plasmon coupling (**Figure 1**). Moreover, AuNPs have very high molar extinction coefficients (ε) (ca. 10^8–10^9 $M^{-1}cm^{-1}$ for AuNPs with diameters between 10 and 50 nm), which confers this sensing method a great sensitivity. In fact, the color change associated to the aggregation process can be observed by naked eye even at nanomolar concentration [3, 4].

The main challenge in the design of colorimetric sensors with AuNPs is the election of the recognition units to be attached onto the surface of the nanoparticles and the nature of the molecular interaction leading to the aggregation process.

A large number of target analytes (metal ions, anions, small organic molecules, or large biomolecules) have been detected using functionalized AuNPs as colorimetric probes. However, in this chapter, the discussion has been limited to the use of functionalized spherical gold nanoparticles for the detection of small molecules with biological activity, such as neurotransmitters, nerve agents, pesticides, and biologically relevant carboxylates.

2. Detection of neurotransmitters

Among the neurotransmitters, biogenic amines (BA) are of particular interest due to their impact in areas ranging from biomarkers of specific diseases [5–7] to quality control of foodstuffs [8, 9]. Nitric oxide (NO) is also very important neurotransmitter in the central, peripheral, and enteric nervous systems (ENS) [10, 11].

Dopamine, the simplest biogenic catecholamine (CA), is an important neurotransmitter of the central and peripheral nervous systems [12]. An approach for the colorimetric detection of dopamine has been developed using 4-amino-3-hydrazino-5-mercapto-1,2,4-triazol functionalized AuNPs (**Figure 2**).

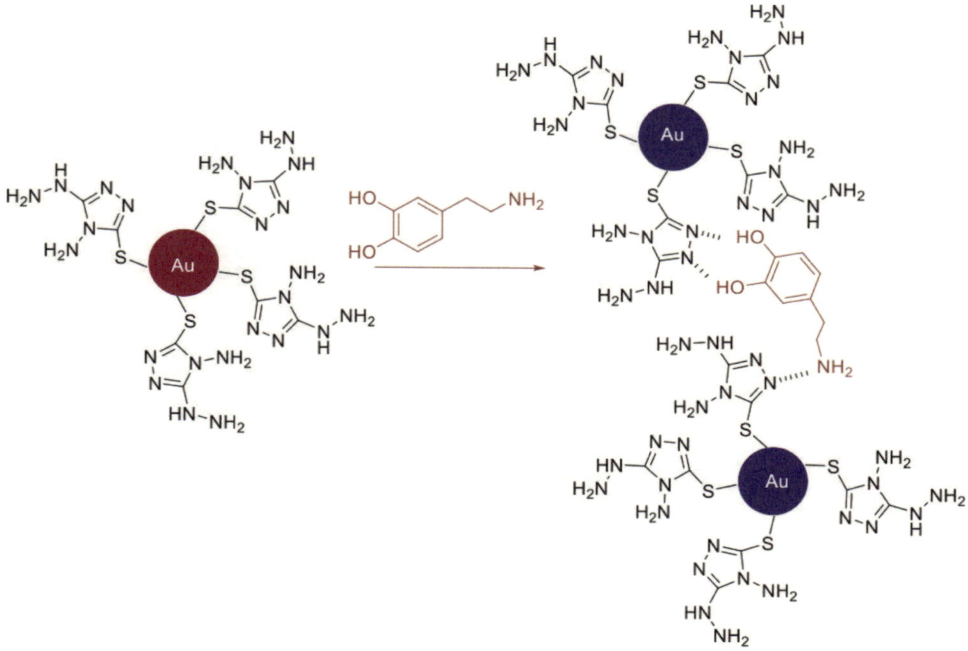

Figure 2.
Dopamine detection using triazol functionalized AuNPs.

Dopamine induced the aggregation of the AuNPs through hydrogen bonding interactions [13]. Each dopamine molecule has three H-donor groups, which are able to form hydrogen bonds (the amino and both hydroxyl groups). On the other hand, the 4-amino-3-hydrazino-5-mercapto-1,2,4-triazol presents two hydrogen bond acceptors that can interact with the target molecule, inducing aggregation with the concomitant color change. Epinephrine, norepinephrine, and isoprenaline were tested as possible interferants. The three compounds showed lower responses than dopamine.

Following the same approach, several functionalized AuNPs have been reported [14–16] for dopamine detection in biological media. This neurotransmitter has also been detected using unmodified citrate-capped gold nanoparticles [17]. A net of hydrogen bonds among dopamine molecules and dopamine with citrate is responsible for the aggregation of the nanoparticles (**Figure 3**). Selectivity toward dopamine is achieved by modifying nanoparticle size.

Serotonin (5-hydroxytryptamine, 5-HT) is a neurotransmitter that plays a key role in the regulation of various biological functions. 5-HT has been used as a biomarker to detect the presence of carcinoid tumors. A selective and sensitive probe based on AuNPs for detecting serotonin has been reported [18]. In this case, AuNPs were bi-functionalized with dithiobis(succinimidyl propionate) (DSP) and with *N*-acetyl-L-cysteine (NALC). DSP reacts with the amino group of 5-HT, and NALC is able to interact with the hydroxyl group of serotonine through electrostatic interactions and hydrogen bonds formation, and additionally, it also acts as a stabilizer of the colloidal solutions of AuNPs due to its negative charge at neutral pH (**Figure 4**). In the presence of the analyte, aggregation takes place and the solution color changes from red to blue, being this change observable by naked eye.

Nitric oxide (NO) is one of the gaseous neurotransmitters. NO is generated in the nitric oxide synthase catalyzed oxidation of L-arginine to L-citrulline and is involved in a variety of important biological processes. For example, it stimulates

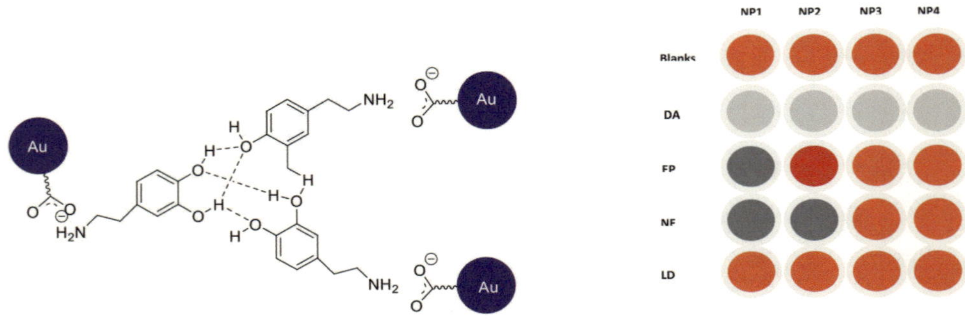

Figure 3.
Mode of interaction and colorimetric response of AuNPs with four different particle sizes against catecholamines (CAs). Experimental conditions: CCAs = 4 μM, CNaCl = 14 mM, pH = 7.0, and incubation time: 10 min. NP1: 13 nm AuNPs; NP2: 23 nm AuNPs; NP3: 32 nm AuNPs; and NP4: 43 nm AuNPs.

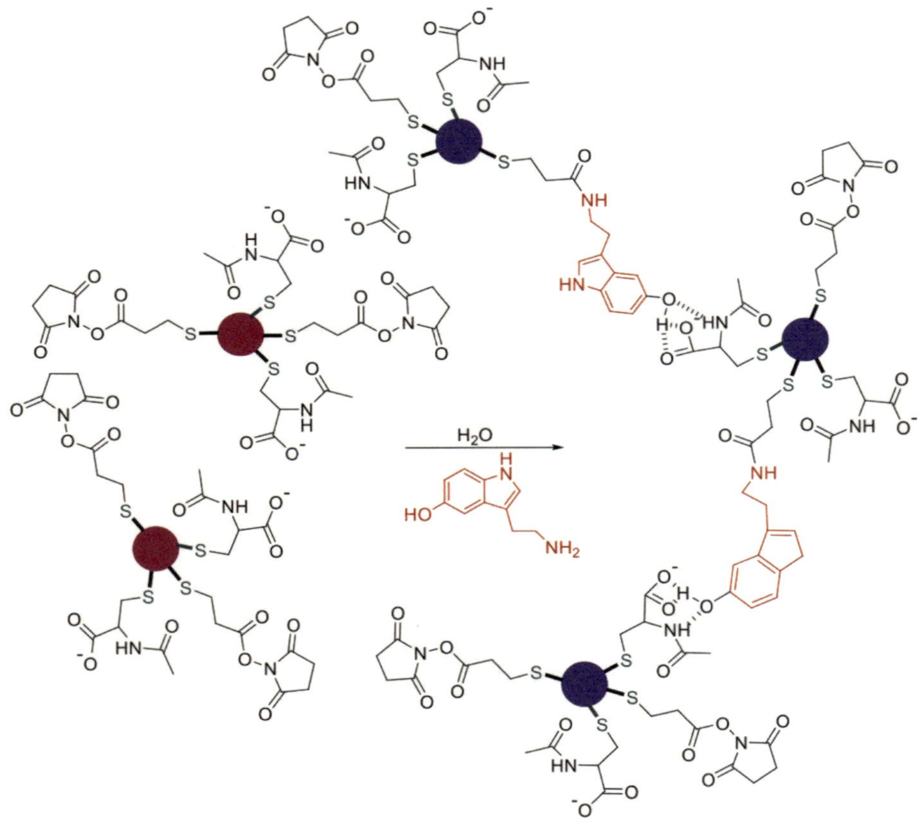

Figure 4.
Detection of serotonin with DSP and NALC functionalized AuNPs.

the production of cyclic guanosine monophosphate (cGMP) which acts as a second messenger. In addition, nitric oxide is released by some neurons that innervate the gastrointestinal tract, penis, respiratory passages, and cerebral blood vessels. Nitric oxide is also released as a neurotransmitter in the brain and has been implicated in the processes of learning and memory. Detection of this gas has been carried out using functionalized AuNPs that aggregates through NO-induced "click" reaction [19]. To detect the analyte, both azide- and terminal alkyne-functionalized gold nanoparticles were synthesized.

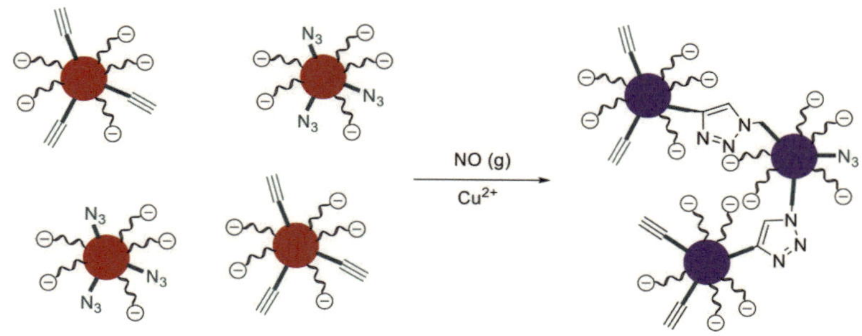

Figure 5.
Mechanism for detecting NO with a mixture of azide and alkyne functionalized AuNPs.

The sensing mechanism is shown in **Figure 5**. The initial aqueous solution containing a mixture of azide- and terminal-alkyne functionalized AuNPs remains dispersed in the presence of Cu(II), with its characteristic red-wine color. When NO is added to the solution, Cu(II) is reduced to Cu(I), and then, the "click" reaction between azide and alkyne-terminated nanoparticles takes place, giving rise to the nanoparticles aggregation with the subsequent change in the color of the solution.

3. Detection of nerve agents and pesticides

The current rise in international concern over the use of chemical warfare (CW) agents in different conflictive sceneries has resulted in an increasing interest in the detection of these lethal chemicals. Among CW species, nerve agents are greatly dangerous and their high toxicity and easy production underscore the need to detect these deadly chemicals via quick and reliable procedures. AuNPs have been used as sensors for some nerve agent simulants with good results. Different sensing mechanisms have been used for detecting these compounds, for example, compensation of charges has been applied for this process [20]. The designed sensing protocol takes advantage of the nucleophilic reactivity of pyridine moieties toward nerve gases [21]. This reactivity generates positive charges on the surface of the gold nanoparticles, diminishing their colloidal stability and inducing aggregation (**Figure 6**).

Several pyridine derivatives were studied, and their ability to act as probes for DCNP detection was evaluated by UV-vis spectroscopy. After addition of increasing amounts of DCNP, the intensity of the surface plasmon peak of the monodispersed AuNPs at 526 nm decreased and a new peak at c.a. 660 nm appeared as the AuNP clusters were formed. The best results were obtained with compound **1** (**Figure 6c**) that showed a limit of detection of 76 ppm under the experimental conditions used.

Following the same approach, compound **2** (**Figure 7**) was bound to gold nanoparticles. In this case, the positive charge appears as a consequence of the reaction described in **Figure 7**. The limit of detection determined in this case was 81 ppm.

Also, triarylcarbinols have been used as recognition motifs. These compounds can be converted into their corresponding carbocations in the presence of nerve agent simulants through phosphorylation of the OH group followed by S_N1 elimination. Consequently, AuNPs functionalized with this type of compounds have been used in detecting simulants of these dangerous compounds (**Figure 8**, X = F, CN) [22].

An approach based on enzymatic reactions has also been described for detecting nerve agents GB, GD, and VX and the highly toxic pesticide paraoxon (**Figure 9**). The prepared sensor showed high sensitivity [23].

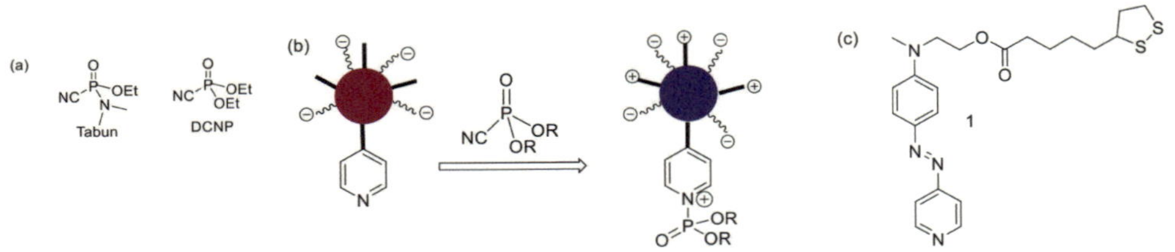

Figure 6.
(a) Nerve agent Tabun and its mimic DCNP (b) Proposed mechanism for the colorimetric detection of DCNP. (c) Pyridine derivative used as recognition unit.

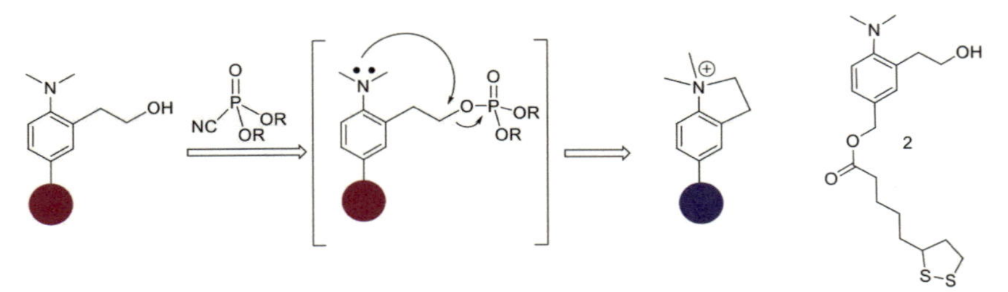

Figure 7.
Mechanism of positive charge generation upon reaction of DCNP with AuNPs functionalized with compound 2.

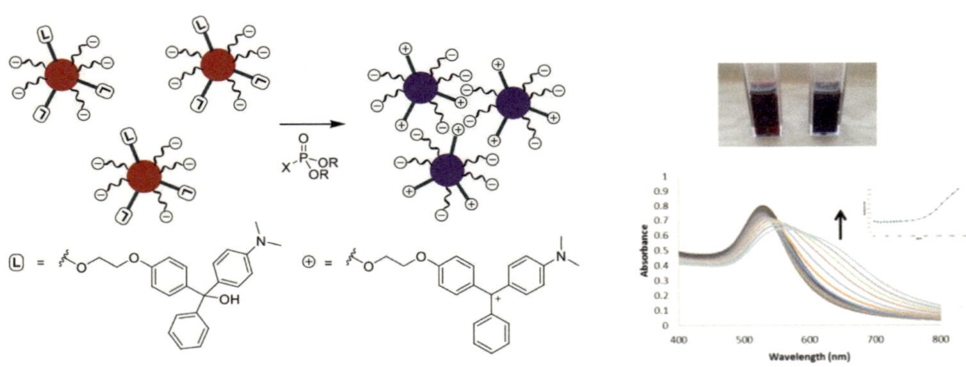

Figure 8.
(Left) AuNPs functionalized with triarylcabinols for detecting nerve agent simulants. (Right) Color change observed on the solution in DMF upon addition of the simulants. UV-vis spectra of the triarylcarbinol functionalized AuNPs on addition of increasing amounts of DCNP expressed mg/m^3. Insets: Plots A_{640}/A_{526} vs. DCNP and DFP concentration, respectively.

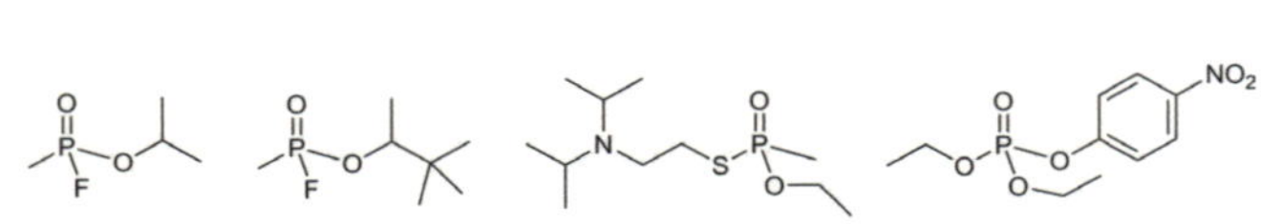

Figure 9.
Structures of nerve agents GB, GD, and VX, and pesticide paraoxon.

In this method, thiocholine (TCh), which is generated from S-acetylthiocholine (ATCh) through acetylcholinesterase (AChE) enzymatic hydrolysis, induces aggregation of AuNPs stabilized with lipoic acid. When the analytes are present in the solution, the production of TCh is suppressed, and a disaggregation of the particles is observed with a change of color from blue to red (**Figure 10**).

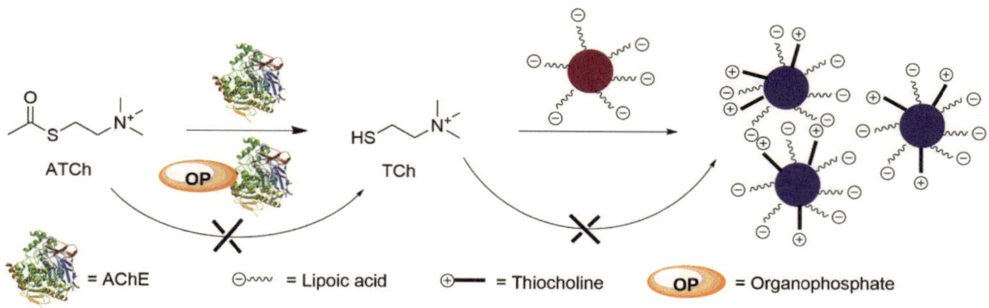

Figure 10.
Sensing mechanism based on AChE inhibition.

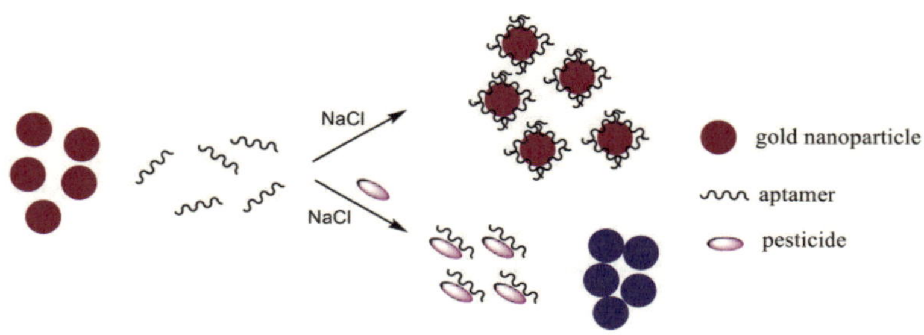

Figure 11.
Sensing mechanism using aptamers.

Neither AChE nor ATCh can induce aggregation of the AuNPs stabilized with lipoic acid, whereas TCh is capable of producing this process. The analytes inactivate AChE via nonreversible phosphorylation even with very low concentrations. In consequence, ATCh cannot be hydrolyzed which inhibits the aggregation. Limits of detection are in the pM range and the system can be used in complex matrices such as in apple juice.

Other dangerous pesticides have been detected using AuNP-based colorimetric aptasensors [24]. The sensing mechanism is summarized in **Figure 11**. Aptamers (organophosphorous pesticide aptamers in this case) are able to be adsorbed on the surface of gold nanoparticles through coordination between Au and N atoms in DNA bases. When salt is added, this adsorption gives rise to stable dispersions of gold nanoparticles with the characteristic red color. However, in the presence of the analytes, the aptamers desorb from the surface of the AuNPs, giving rise to the aggregation process with the concomitant change of color. Because double-stranded DNA is not adsorbed by gold, single-stranded DNA was used in these experiments. Although the sensitivity of these assays is not high enough, this design shows a new approach for preparing sensors based on gold nanoparticles.

4. Detection of biologically important carboxylic acids

Detection of dicarboxylates is very important, since some of them have critical roles in the most important metabolic cycles of living organisms. Specifically, succinate plays a fundamental role in processes, such as the Krebs cycle and oxidative phosphorylation. It is an inhibitor of mitochondrial lipid peroxidation, preventing or delaying most of the damage caused by the peroxidation on different mitochondrial structures and functions. For all these reasons, it is of interest to detect this compound. Among the different strategies used in designing sensors for these compounds, functionalized gold nanoparticles have shown interesting applications [25].

The sensing protocol, in this case, is related to the well-known interaction between the carboxylate and thioureas groups. For this reason, AuNPs containing thiourea groups were prepared, and their behavior in front of different dicarboxylates was evaluated. Among the different dicarboxylates studied (oxalate (**C2**), malonate (**C3**), succinate (**C4**), glutarate (**C5**), and adipate (**C6**); all of them as TBA salts) only with succinate, a color change of the solution from red to blue was observed (**Figure 12**, right, bottom). The limit of detection determined for this dianion was 0.5 μM.

Following a similar approach, maleate and fumarate were distinguished [26, 27]. Compound **3** was used to cap gold nanoparticles (**Figure 13**). The prepared AuNPs were able to recognize trans-dicarboxylates such as fumarate, one of the key components generated in the Krebs cycle, over its cis-isomer, maleate. The trans-isomer, fumarate, seems to have the appropriate geometry to induce the nanoparticle aggregation, whereas the cis compound presents a similar behavior to that of flexible dicarboxylates like oxalate, malonate, succinate, glutarate, propionate, and 4-pentenoic acid.

Pyruvic acid (2-oxopropanoic acid, PA) is the simplest alpha-keto acid. It plays important roles in several biochemical pathways. For example, it supplies energy to cells through the citric acid cycle when oxygen is present and under hypoxic conditions produces lactate. It also appears as an intermediate in several metabolic processes such as the glycolysis of glucose or the synthesis of carbohydrates or fatty acids. Gold nanoparticles have proved to be useful for detecting this acid [28]. The approach, in this case, is also based on the use of unmodified AuNPs and Cytidine-rich oligonucleotides (C-rich DNA). C-rich DNA can fold into one closely packed four-stranded structure called i-motif [29] through protonated cytosine-cytosine (C-C+) base-pair formation under slightly acidic conditions. The principle of the designed PA sensor is similar to that previously indicated in **Figure 11**. The rigid i-motif structure

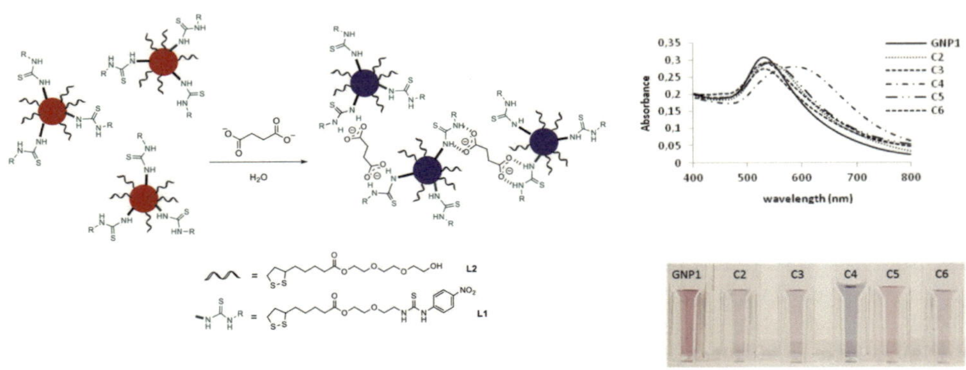

Figure 12.
Detection of succinate.

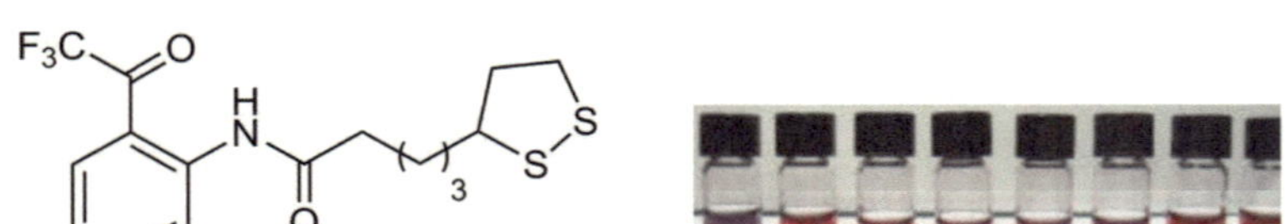

3

Figure 13.
(Left) Ligand used to cap gold nanoparticles for detection of fumarate. (Right) Color changes of an aqueous solution of AuNPs capped with 3 *upon addition of various analytes (as sodium salt). From the left: fumarate, maleate, oxalate, malonate, succinate, glutarate, propionate, and 4-pentenoic acid.*

is unable to stabilize gold nanoparticles, but when a change of pH is induced in the medium from acid to neutral, the C-rich DNA changes its structure and prevents gold nanoparticles aggregation. The change in pH is induced by the addition of PA and pyruvate decarboxylase (PDC). The enzyme transforms PA into acetaldehyde and CO with the corresponding change of pH from acid to neutral. At neutral pH, the C-rich DNA presents its extended single-stranded structure and effectively binds to AuNPs, stabilizing them against NaCl-induced aggregation. Based on this principle, PA can be selectively detected by the color change of the AuNPs (**Figure 14**).

UV-Vis spectra were recorded to demonstrate the proposed sensing mechanism. Solutions of AuNPs (3 nM) containing C-rich DNA showed a maximum absorption peak centered at 520 nm. After addition of PA, the changes induced in the absorption band depend on the acid concentration. At 5.6 mM, a hypochromic effect was observed, but at 16.8 mM, a bathochromic shift can be observed from 520 to over 600 nm with the corresponding change of color. After addition of PCD, the disaggregation is produced and the solution recovers its red color. Selectivity against lactic acid, ascorbic acid, and glucose was established, and the limit of detection determined was 3.0 mM.

Ascorbic acid (AA, also known as vitamin C), an antioxidant compound, is present not only in biological fluids but also in foodstuffs and pharmaceuticals. Taking into account the red-ox properties of ascorbic acid, an approach using gold nanoparticles and Cr(VI) has been described for its detection [30]. Gold nanoparticles were stabilized with sodium tripolyphosphate ($Na_5P_3O_{10}$), and the sensor was prepared using this AuNPs and Cr (VI) salts. As Cr(VI) exists in the form of $HCrO_4^-$, CrO_4^{2-}, or $Cr_2O_7^{2-}$, there is an electrical repulsion with tripolyphosphate that precludes the nanoparticles aggregation. By contrast, Cr(III), a hard lewis acid strongly coordinates to the polyphosphate ligand, giving rise to the charges compensation and in consequence the nanoparticles aggregation (**Figure 15**). Ascorbic acid is able to trigger the process by reducing Cr(VI) to Cr(III). Selectivity studies were carried out with PO_4^{3-}, Zn^{2+}, Cu^{2+}, SO_4^{2-}, Ni^{2+}, Li^+, Na^+, Mg^{2+}, Br^-, NO_3^-, glucose (Glu), citric acid (CA), and oxalic acid (OA) as interferences, and no appreciable changes were observed with any of the studied compounds. The limit of detection determined for this method was 0.15 μM.

Finally, also amino acids have been detected using functionalized gold nanopaticles. Thus, tyrosine (Tyr) was detected using N-acetyl-L-cysteine modified gold nanoparticles [31]. In this case, the chiral ligand N-acetyl-L-cysteine (NALC) was chosen to include chirality in the sensors and study its use in selective recognition and separation of enantiomers.

The chiral selectivity is attributed to a chemical interaction between chiral NALC–Au NPs and L-Tyr at the molecular level as is shown in **Figure 16**.

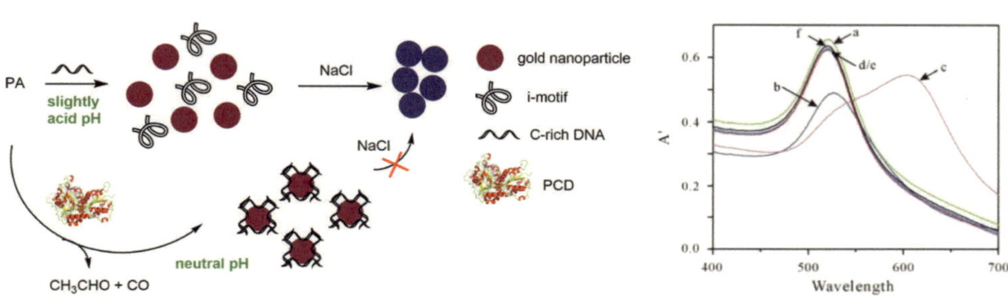

Figure 14.
(Left) Mechanism for pyruvic acid detection. (Right) UV-Vis spectra of the AuNP suspensions (3 nM) containing C-rich DNA after incubation with (b) 5.6 mM PA, (c) 16.8 mM PA, (d) 5.6 mM PA + PDC, (e) 16.8 mM PA + PDC, and (f) PDC only; curve (a) is the background signal (reproduced with license number Reprinted with permission from Li et al. [28]. Copyright 2014 Royal Society of Chemistry).

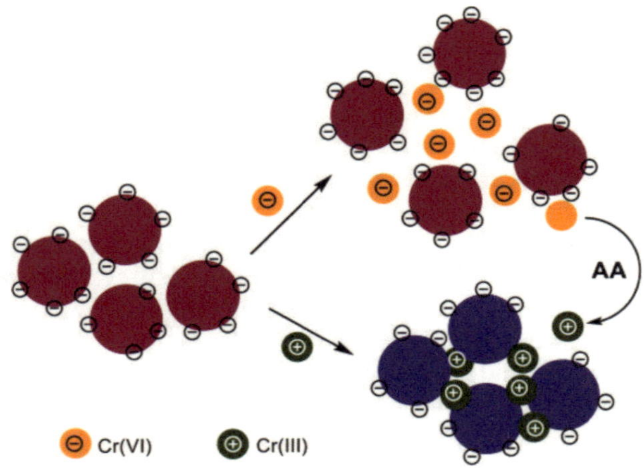

Figure 15.
Detection of ascorbic acid (AA).

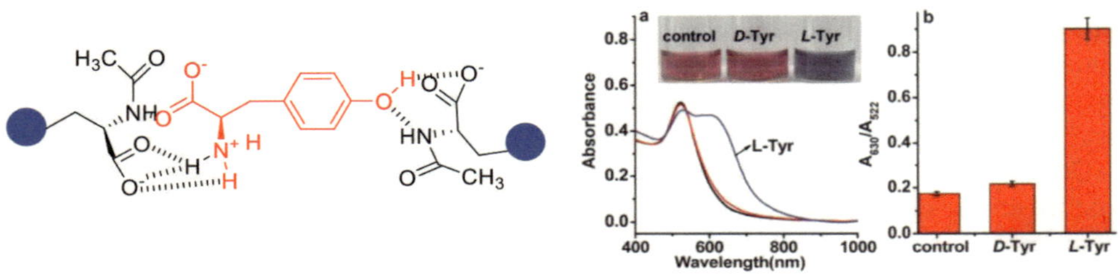

Figure 16.
(Left) Interaction between tyrosine and N-acetyl-L-cystein capped gold nanoparticles. (Right) Chyral selectivity showed by the sensor. Reprinted with permission from Su et al. [31]. Copyright 2014 Royal Society of Chemistry.

Tyr interacts with the ligand NALC through hydrogen bonds that involve carboxylic, amino, and hydroxyl groups. The selectivity response may be attributed to the conformation of L-Tyr that seems to be more appropriate for forming the complex with NALC.

5. Conclusions

The use of gold nanoparticles for the preparation of colorimetric sensors is a very active field. The changes in the color of colloidal gold nanoparticles in solution because of the change in the surface plasmon absorption band upon aggregation or disaggregation processes can be easily used to transform the molecular recognition event into a macroscopic measurable signal. This change from red to blue can be observed by the nacked eye, allowing in this way cheap and easy detection of the target analytes. In this chapter, different mechanisms for the direct or indirect analyte-triggered aggregation of the AuNPs have been considered, including chemical reactions, supramolecular interactions, or changes in the pH of the medium. The selectivity observed in the sensing response in some cases depends on the conformation or configuration of the analyte, but also it can be achieved by using enzymes that catalyze specific reactions or aptamers able to interact with an analyte. Reactions induced by the analyte have also been explored, for example based on redox transformations. Taking into account the interesting photophysical properties of the gold nanoparticles, their easy functionalization, the use of aqueous solutions, and the detection by naked eye, we can conclude that the red or blue question will continue to be very present in the molecular sensing field.

Author details

Pablo Gaviña*, Margarita Parra, Salvador Gil and Ana M. Costero
Instituto Interuniversitario de Reconocimiento Molecular y Desarrollo Tecnológico (IDM), Universitat de València—Universitat Politècnica de València, Valencia, Spain

*Address all correspondence to: pablo.gavina@uv.es

References

[1] Saha K, Agasti SS, Kim C, Li X, Rotello VM. Gold nanoparticles in chemical and biological sensing. Chemical Reviews. 2012;**112**:2739-2779. DOI: 10.1021/cr2001178

[2] Mayer KM, Hafner JH. Localized surface plasmon resonance sensors. Chemical Reviews. 2011;**111**:3828-3857. DOI: 10.1021/cr100313v

[3] Bedford EE, Spadovecchia J, Pradier CM, Gu FX. Surface Plasmon resonance biosensors incorporating gold nanoparticles. Macromolecular Bioscience. 2012;**12**:724-739. DOI: 10.1002/mabi.201100435

[4] Liu Y, Liu Y, Liang Z, Li X, Liu S, Yu J. Enhanced sensitivity and selectivity of chemosensor for malonate by anchoring on gold nanoparticles. Chinese Journal of Chemistry. 2011;**29**:531-538. DOI: 10.1002/cjoc.201190118

[5] Marecos C, Ng J, Kurian MA. What is new for monoamine neurotransmitter disorders? Journal of Inherited Metabolic Disease. 2014;**37**:619-626. DOI: 10.1007/s10545-014-9697-4

[6] Plonka J. Methods of biological fluids sample preparation—biogenic amines, methylxanthines, water-soluble vitamins. Biomedical Chromatography. 2015;**29**:1-10. DOI: 10.1002/bmc.3353

[7] Bachrach U. Polyamines and cancer: Minireview article. Amino Acids. 2004;**26**:307-309. DOI: 10.1007/s00726-004-0076-6

[8] Jairath G, Singh PK, Dabur RS, Rani M, Chaudhari M. Biogenic amines in meat and meat products and its public health significance: a review. Journal of Food Science and Technology. 2015;**52**:6835-6846. DOI: 10.1007/s13197-015-1860-x

[9] Guo Y-Y, Yang YP, Peng Q, Hang Y. Biogenic amines in wine: A review. International Journal of Food Science and Technology. 2015;**50**:1523-1532. DOI: 10.1111/ijfs.12833

[10] Goyal RK, Hirano I. The enteric nervous system. The New England Journal of Medicine. 1996;**334**:1106-1115. DOI: 10.1056/NEJM199604253341707

[11] Stark ME, Szurszewski JH. Role of nitric oxide in gastrointestinal and hepatic function and disease. Gastroenterology. 1992;**103**:1928-1949. PMID: 1333429

[12] Adams RN. Probing brain chemistry with electroanalytical techniques. Analytical Chemistry. 1976;**48**:1126A-1138A. DOI: 10.1021/ac50008a001

[13] Feng JJ, Guo H, Li YF, Wang YH, Chen WY, Wang AJ. Single molecular functionalized gold nanoparticles for hydrogen-bonding recognition and colorimetric detection of dopamine with high sensitivity and selectivity. ACS Applied Materials and Interfaces. 2013;**5**:1226-1231. DOI: 10.1021/am400402c

[14] Chen Z, Zhang C, Wang CA. Colorimetric assay of dopamine utilizing melamine modified gold nanoparticle probes. Analytical Methods. 2015;**7**:838-841. DOI: 10.1039/c4ay02564d

[15] Saravanan G, Ankireddy SR, Viswanath B, Kim J, Yun K. Fluorescent gold nanoclusters for selective detection of dopamine in cerebrospinal fluid. Scientific Reports. 2017;7:4029. DOI: 10.1038/srep40298

[16] Kong B, Zhu A, Luo Y, Tian Y, Yu Y, Shi G. Sensitive and selective colorimetric visualization of

cerebral dopamine based on double molecular recognition. Angewandte Chemie International Edition. 2011;**50**: 1837-1840. DOI: 10.1002/anie.20100707

[17] Mohseni N, Bahram M. Highly selective and sensitive determination of dopamine in biological samples via tuning the particle size of label-free gold nanoparticles. Spectrochimica Acta Part A: Molecular and Biomolecular Spectroscopy. 2018;**93**:451-457. DOI: 10.1016/j.saa.2017.12.033

[18] Godoy-Reyes TM, Llopis-Lorente A, Gaviña P, Costero AM, Martínez-Máñez R, Sancenón F. Selective and sensitive colorimetric detection of the neurotransmitter serotonin based on bi-functionalized gold nanoparticles. Sensors and Actuators, B: Chemical. 2018;**258**:829-835. DOI: 10.1016/j.snb.2017.11.181

[19] Martí A, Costero AM, Gaviña P, Parra M. Selective colorimetric NO(g) detection based on the use of modified gold nanoparticles using click chemistry. Chemical Communications. 2015;**51**:3077-3079. DOI: 10.1039/c4cc10149a

[20] Martí A, Costero AM, Gaviña P, Gil S, Parra M, Brotons-Gisbert M, Sánchez-Royo JF. Functionalized gold nanoparticles as an approach to the direct colorimetric detection of DCNP nerve agent simulant. European Journal of Organic Chemistry. 2013;**2013**:4770-4779. DOI: 10.1002/ejoc.201300339

[21] Royo S, Costero AM, Parra M, Gil S, Martinez-Mañez R, Sancenón F. Chromogenic, specific detection of the nerve-agent mimic DCNP (a Tabun Mimic). Chemistry—A European Journal. 2011;**17**:6931-6934. DOI: 10.1002/ejoc.201300339

[22] Martí A, Costero AM, Gaviña P, Parra M. Triarylcarbinol functionalized gold nanoparticles for the colorimetric detection of nerve agent simulants. Tetrahedron Letters. 2014;**55**:3093-3096. DOI: 10.1016/j.tetlet.2014.03.139

[23] Sun J, Guo L, Bao Y, Xie J. A simple, label-free AuNPs-based colorimetric ultrasensitive detection of nerve agents and highly toxic organophosphate pesticide. Biosensors and Bioelectronics. 2011;**28**:152-157. DOI: 10.1016/j.bios.2011.07.012

[24] Bai W, Zhu C, Liu J, Yan M, Yang S, Chen A. Gold nanoparticle–based colorimetric aptasensor for rapid detection of six organophosphorous pesticides. Environmental Toxicology and Chemistry. 2015;**34**:2244-2249. DOI: 10.1002/etc.3088

[25] Martí A, Costero AM, Gaviña P, Parra M. Selective recognition and sensing of succinate vs. other aliphatic dicarboxylates by thiourea-functionalized gold nanoparticles. Chemistry Select. 2016;**5**:1057-1060. DOI: 10.1002/slct.201600313

[26] Youk KS, Kim KM, Chatterjee A, Ahn KH. Selective recognition of fumarate from maleate with a gold nanoparticle-based colorimetric sensing system. Tetrahedron Letters. 2008;**49**:3652-3655. DOI: 10.1016/j.tetlet.2008.03.142

[27] Chatterjee A, Oh DJ, Kim KM, Youk KS, Ahn KH. Selective colorimetric sensing of geometrical isomers of dicarboxylates in water by using functionalized gold nanoparticles. Chemistry—An Asian Journal. 2008;**3**:1962-1967. DOI: 10.1002/asia.200800233

[28] Li W, Pan C, Hou T, Wang X, Li F. Selective and colorimetric detection of pyruvic acid using conformational switch of i-motif DNA and unmodified gold nanoparticles. Analytical Methods. 2014;**6**:1645-1649. DOI: 10.1039/c3ay41883a

[29] Protopopova AD, Tsvetkov VB, Varizhuk AM, Barinov NA, Podgorsky VV, Klinov DV, Pozmogova GE. The structural diversity of C-rich DNA aggregates: Unusual self-assembly of beetle-like nanostructures. Physical Chemistry Chemical Physics. 2018;**20**:3543-3553. DOI: 10.1039/c7cp05380k

[30] Sui N, Liu F, Li T, Wang L, Wang T, Liu M, YU WW. Colorimetric detection of ascorbic acid based on the trigger of gold nanoparticles aggregation by Cr(III) reduced from Cr(VI). Analytical Sciences. 2017;**33**:963-967. DOI: 10.2116/analsci.33.963

[31] Su H, Zheng Q, Li H. Colorimetric detection and separation of chiral tyrosine based on N-acetyl-L-cysteine modified gold nanoparticles. Journal of Materials Chemistry. 2012;**22**:6546-6548. DOI: 10.1039/c2jm16746h

Chapter 3

Silica-Supported Gold Nanocatalyst for CO Oxidation

Ziyauddin S. Qureshi and EA Jaseer

Abstract

Even though gold is inert in its bulk practice, greatly disseminated gold nanoparticles (Au NPs) with dimensions less than 5 nm have been found to be active for a number of oxygen transfer reactions, particularly for low-temperature CO oxidation. The catalytic activity not only be subject to the particle size of Au but also on the nature of the support and the synthesis method of the catalyst. These factors are frequently inter-related such that their separate contributions cannot be easily unraveled. Also, the activity of a supported Au catalyst is ruled by a complex combination of contributions of the particle morphology, metal dispersion, and electronic properties of the gold. Higher catalytic activity is being observed for Au NPs supported on reducible metal oxides such as TiO_2, Co_3O_4, CeO_2, and Fe_2O_3. However, silica is an inert, inexpensive, and convenient support that can be shaped into a host of attractive and varied morphologies. In this chapter, the study of CO oxidation catalyzed by mono- and bimetallic Au NPs over various silica supports is discussed in detail.

Keywords: gold nanoparticles, silica supports, mono-/bimetallic gold nanoparticles, CO oxidation, heterogeneous catalysis

1. Introduction

Since the revelation of the earliest inspiring studies [1], catalysis by Au NPs has been in the limelight of chemical research and it has been recognized as a vital tool for a wide range of transformations such as selective oxidations and hydrogenations of organic substrates [2–6], water-gas shift reaction [7–9], acetylene hydrochlorination [10, 11], direct synthesis of hydrogen peroxide [12], reduction of NO to N_2 [13, 14], and the addition of nucleophiles to acetylenes [15]. Particularly, the aerobic oxidation of CO by gold [16] has been a subject of constant exploration, where it plays a crucial role for industry and the environment [17].

Carbon monoxide is a colorless and odorless gas [18]. Due to its high affinity with hemoglobin, it is enormously toxic for humans and animals. Its long-term exposure should not exceed 25 ppm in 8 h, where more and more obvious harmful effects are observed above this limit up to the lethal concentration of around 650–700 ppm. Massive amount of CO is continuously being emitted (1.09 billion tons in 2000), frequently from transportation, power plants, and industrial and domestic doings [19]. CO is also an originator of ground-level ozone, which can cause several severe respiratory problems. Also, it is not well soluble in water (23 $mL_{CO(g)}\ L^{-1}{}_{H2O(l)}$), which also limits its removal from air by means of aqueous treatments. So its oxidation into CO_2 is a key solution for CO reduction in air depollution managements. To achieve CO elimination, it is necessary to design catalysts, which permits the oxidation reaction with a sufficient rate.

Au NPs supported over metal oxides demonstrate high catalytic activity for CO oxidation under mild conditions and even below room temperature [20]. Concerning the structure-activity relationship, Au NPs size has been witnessed to play a key role in this transformation, with a finest activity observed for gold clusters in the size range of 2–5 nm [21]. The nature of the support and the way the Au NPs is being prepared and interacted are significant [22]. Till the beginning of the twentieth century, scarce oxides such as hopcalite have been identified to oxidize CO at ambient temperature. Hopcalite is composed of a binary amorphous Mn–Cu oxide, whose texture will get stabilized by the addition of CaO, Al_2O_3, or bentonite clay [23]. Later, several oxides have been investigated and higher catalytic activity are being observed for Au NPs supported on various reducible metal oxides such as TiO_2, [24] Co_3O_4, [25] CeO_2, [26], and Fe_2O_3 [27]. However, silica is an inert, inexpensive, and convenient support that can be shaped into a host of attractive and varied morphologies [28]. Silica-based mesoporous supports have been acclaimed in precise for high surface area, high accessibility of the catalytically active center, and reduced particle sintering [29]. However, it is still challenging to make properly sized, well-dispersed, and uniform Au NPs placed within the nanopores of the material without conceding mass transfer and morphological properties [30]. Gold alloy (bimetallic nanocatalysts) along with other metals such as Pd, Cu, and Ag, also found to exhibit high and exceptional catalytic activity compared to Au alone.

There are few reviews on different facets of gold catalysis [31–37]. This chapter will offer an overview of the different possible methodologies available for the synthesis of silica supported especially mono and bimetallic gold nanocatalyst for CO oxidation. The key advantages or limits for the proposed methods are shortly discussed; in some cases, the most important features of the presented methodologies are highlighted from an industrial point of view.

1.1 Methods to synthesize gold nanocatalyst supported on silica for CO oxidation

1.1.1 Post-synthetic treatment of silica before gold loading

Schuth et al. reported the method for the preparation of an active gold catalyst for CO oxidation, supported on silica, made by a novel solution technique [38]. The surface of SBA-15 was functionalized by treating with positively charged groups, and $[AuCl_4]^-$ species into the channel structure through ion exchange. Later on reduction with $NaBH_4$, resulted a highly dispersed Au NPs in the channels system of the mesoporous host. This composite material showed a reaction rate of 2.7×10^{-4} $mmolg^{-1}cats^{-1}$ for CO oxidation. The Au NPs to some extent get stabilized by the pore system of SBA-15. Still, the interaction is relatively weak and sintering was observed around 100°C, which was confirmed from TEM images of Au/SBA-15. The authors conclude that a specific interaction between gold and the support is not compulsory for the generation of very active gold-based catalysts.

Datye et al. demonstrated a novel silica geometry, which allows to trap nanoparticles and limit the mobility of species that lead to thermal sintering [39]. Mesoporous silica was prepared by two different methods, spontaneous self-assembly of amphiphilic molecules (batch synthesis) and evaporation-induced self-assembly (aerosol synthesis). To deposit gold inside the pores of mesoporous silica, firstly silica surface was functionalized with organic amine then treated with Au precursor. These reduced catalysts were tested for CO oxidation with 20% CO and 10% O_2 from room temperature up to 400°C. Results showed that the hexagonal structure inside aerosol silica shells support to contain the Au NPs within the pore structure and protect the active part from gas phase poisons, allowing facile access to gas phase species as related to straight pores in MCM-41.

Datye et al. further investigated the role of pore size and structure in controlling the thermal sintering of Au NPs on mesoporous silica [40]. The phenomenon of sintering of Au NPs is dependent on pore size, pore wall thickness (strength of pores), and pore connectivity. Au was placed on mesoporous silica samples with a varied range of pore sizes and pore arrangements (2-D hexagonal, 3-D hexagonal, and cubic) (**Figure 1**). Later, all catalysts were reduced at 200°C in flow of H_2 and then used for CO oxidation at temperatures ranging from 25 to 400°C. Among all samples, SBA-15 with the thickest pore walls was found most active at regulatory growth of the Au particles.

Mou et al. have prepared the Au NPs embedded within the mesoporous silica's and used as catalysts for CO oxidation [41]. The silane APTS ($H_2N(CH_2)_3$-$Si(OMe)_3$) was used for the surface-functionalization of mesoporous silica such as MCM-41, MCM-48, and SBA-15. The catalysts were activated by calcinations and later reduced with H_2 reduction at 600°C. The catalysts were found to be active for CO oxidation, reaching a very high activity values as 7.0×10^{-3} mmolg^{-1}cats^{-1} at

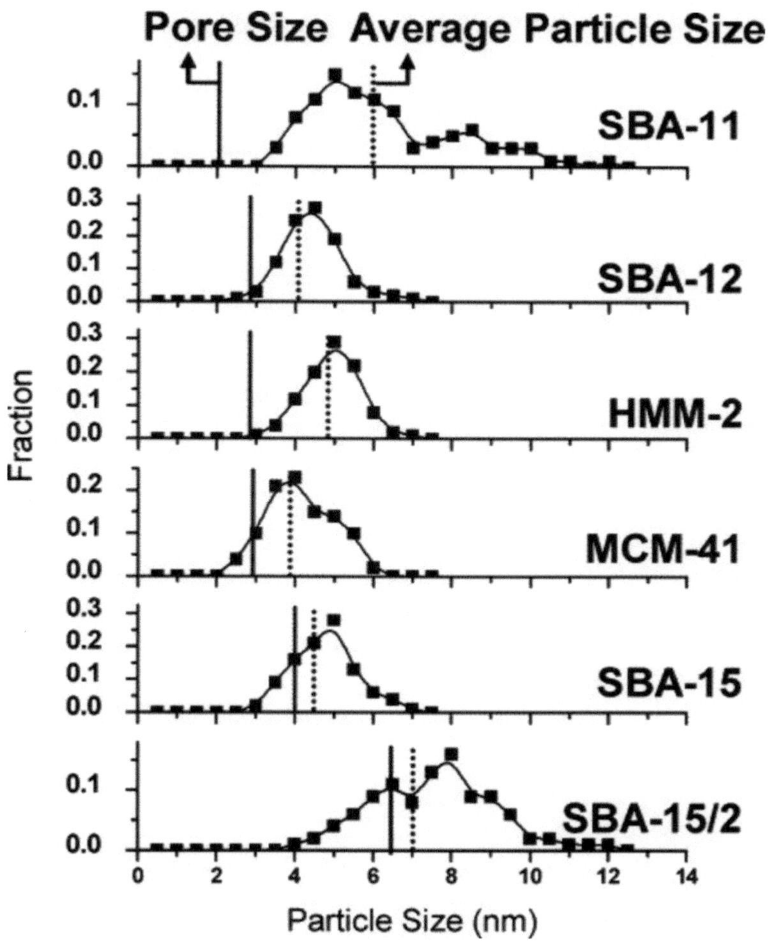

Figure 1.
Au particle size distributions after reduction at 200°C for 2 h in flowing hydrogen. Pore sizes of the samples are calculated using BJH theory. Average Au particle sizes are number averages. Reproduced with permission, copyright 2018, American Chemical Society.

80°C as compared to the results from catalysts made with the standard precipitation-deposition method. Studies also recommend that the CO conversion increases with decreasing size of Au NPs.

Rombi et al. [42] have functionalized SBA-15 with new mercaptopropyltrimethoxysilane (MPS) and used for the preparation of supported gold nanocatalyst for the low-temperature CO oxidation reaction. It has been observed that, the presence of organic residues and the size of the Au NPs intensely affect catalytic activity. High temperature calcination (300–560°C) in air followed by treatment (600°C) under H_2 atmosphere only leads to the formation of homogeneously small size Au spherical nanocrystals dispersed inside the SBA-15 channels with average crystal sizes of about 6.5 nm. The catalysts were first calcined at 560°C and then reduced under H_2/He 600°C. Though, regardless of their large particle size, remarkable catalytic activity for CO oxidation was observed, while the as-made catalysts were found to be inactive for CO oxidation at low temperatures. Gold was found mostly in metallic Au(0) state, due to the strong interaction between the mercapto functional groups and Au, positively charged $Au^{\delta+}$ species also exist.

SBA-15 functionalization with 3-MPS has been used by Ferino et al. to make a supported gold catalyst for the low-temperature CO oxidation [43]. Catalytic runs were studied by varying different reaction parameters such as atmospheric pressure, temperature 40–150°C, and different thermal treatments of the sample prior to reaction. The catalyst was pretreated toward different thermal treatments, that is, calcination in air, treatment in H_2 atmosphere, and calcination followed by H_2 treatment. It has been observed that, the pre-treatment conditions sturdily affect not only the gold particle size but also the nature of the Au surface species. A substantial catalytic activity for CO oxidation was observed for the catalysts treated at 600°C in H_2/He atmosphere, only after the removal of functionalizing agent by an earlier high-temperature calcination (**Table 1**).

Rombi et al. further slightly modified the procedure for the preparation of silica-supported gold catalyst for CO oxidation by functionalizing the silica surface with 3-MPS, anchoring gold using $HAuCl_4$ solution, and later reducing it with sodium citrate [44]. Before the catalytic runs, the Au/SiO_2–SH catalyst was submitted to different thermal treatments. It has been achieved a notable CO conversion when the catalyst was calcined in air at 560°C and afterward treated in H_2/He at 600°C or directly treated in H_2/He at 600°C.

1.1.2 Direct synthesis

Lin et al. have studied CO oxidation over Au/SiO_2 and Au/TiO_2 catalysts [45]. It has been found that, after a high-temperature reduction (HTR) at 500°C, TiO_2-supported Au became very active for CO oxidation at 40°C, an order of magnitude

Thermal treatment	Conversion (mol%)		TOF (s-1)	
	40 °C	80 °C	40 °C	80 °C
Air 560 °C	1.85	1.96	$1.11. 10^{-2}$	$1.17. 10^{-2}$
H_2/He 600 °C	0.55	0.87	$1.57. 10^{-3}$	$2.46. 10^{-3}$
Air 560 °C–H_2/He 600 °C	11.60	22.74	$7.80. 10^{-2}$	$1.53. 10^{-1}$
H_2/He 600 °C –air 560 °C –H_2/He 600 °C	21.13	31.47	$8.21. 10^{-2}$	$1.22. 10^{-1}$

Table 1.
Catalytic performance of the Au/SBA-15-SH sample for the low-temperature CO oxidation reaction.

more active than Au/SiO_2. However a low-temperature reduction (LTR) at 200°C made an Au/TiO_2 catalyst with very low activity. It has been observed that, a HTR step followed by calcination at 400°C and a LTR step (HTR/C/LTR) provided the most active Au/TiO_2 catalyst of all, which at 40°C 100-fold more active than a typical 2% Pd/Al_2O_3 catalyst and also found to be stable above temperature 127°C, while a sharp reduction in activity was observed with the other Au/TiO_2 (HTR) sample. Besides, the activity of the Au/SiO_2 catalyst was ten-fold higher than the Au/TiO_2 (LTR) sample but ten-fold lower than the Au/TiO_2 (HTR/C/LTR) catalyst. Later, Cl analysis disclosed that the Au/TiO_2 (LTR) sample retained around 50% of the Cl from the Au precursor, while the other three catalysts retained only 16%, which gave a final Cl/Au ratio of 0.5; subsequently, the low activity of the Au/TiO_2 (LTR) sample may be due to its high chloride content. However, the inhibitive role of Cl impurity has not yet been established.

Au catalysts supported over mesoporous silica and titania supports were made by Overbury et al. and tested for the oxidation of CO [46]. Two methods were used for the preparation of silica-supported catalysts using triamine as complexing ligands, which lead to mesoporous silica with wormhole and hexagonal structures. The use of triamine ligands is the essential for the synthesis of uniformly sized 2–3 nm Au MPs in the silica pores. While over mesoporous titania, high gold dispersions were obtained without using functional ligand. It was noticed that, Au supported on titania showed a much higher activity for CO oxidation, even the Au particle sizes were essentially equal on the titania and the wormhole silica support. The results advise that the presence of small size Au NPs (2–3 nm) alone is not very satisfactory to achieve high activity in CO oxidation (**Figure 2**). Instead, the support also impacts the activity through other possible ways such as stabilization of active sub-nanometer particles, formation of active oxygen-containing reactant intermediates (such as hydroxyls or O_2), or stabilization of optimal Au structures.

Effect of moisture on the catalytic activity for CO oxidation over three gold catalysts prepared over TiO_2, Al_2O_3, and SiO_2 supports were investigated by Date et al. [47]. A varied range of concentrations was studied, from about 0.1 to 6000 ppm H_2O.

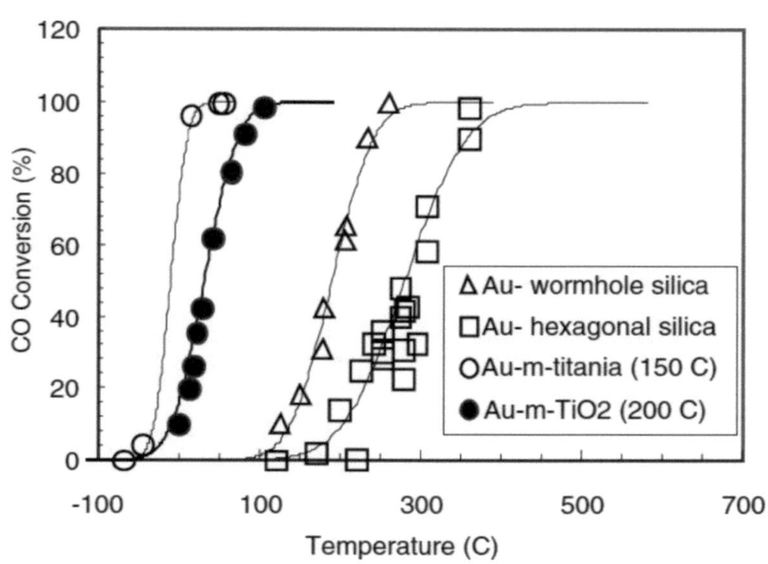

Figure 2.
CO conversion (points) is shown as function of temperature for different supported Au catalysts. Reproduced with permission, copyright 2018, Springer.

It has been observed that, the degree of rate of improvement rest on the type of support used: high for insulating Al_2O_3 and SiO_2 and moderate for semiconducting TiO_2. The effect of moisture becomes notable only above about 200 ppm H_2O for Au/Al_2O_3, while the activity for Au/SiO_2 decreases significantly with a decrease in moisture to about 0.3 ppm. While the catalytic activity of Au/TiO_2 at about 3000 ppm H_2O is very high and reaches full CO conversion (100%). The authors also proposed as a mechanism model that moisture improves the catalytic activities for no less than two orders of magnitude and the effect of moisture depends over the type of metal oxides used. Moisture plays dual role in the reaction: one is the activation of oxygen and other is the decomposition of carbonate.

Dai et al. demonstrated the surface sol-gel process (SSP) for the modification of silica mesopores surfaces and the tuning of mesopore diameters [48]. The procedure involves the preparation of one or multilayers of titanium oxide over SBA-15 [49]. This layer-by layer approach controls the mesopore diameters with monolayer precision (**Figure 3**). Ultra-small Au NPs were effectively prepared on surface-modified SBA-15 via a deposition-precipitation method without the restriction of surface isoelectric point (**Figure 3**). These developed materials are found very active catalysts for CO oxidation.

Corma et al. presented a modified sol-gel approach for the preparation of Au NPs embedded in silica, and Au NPs are capped with two different thiol molecules, 1-dodecanethiol (DT) and 3-mercaptopropyltrimethoxysilane (MPMS), which lead to high-surface area Au/SiO_2 catalysts [50]. The synthesis comprises the formation of a three-component metal-organic-inorganic structure collected of Au NPs capped with alkanethiols and partly functionalized with alkoxysilane groups, and polymerized with tetraethyl orthosilicate (TEOS). Upon calcination, the material becomes highly active catalyst for the CO oxidation at low temperatures, the same as that obtained with gold on TiO_2. This clearly specifies that it is possible to attain highly active gold-silica catalysts by a liquid-based method, as long as accessible small gold particles and strong metal-support interactions exist.

The structure and oxidation state of gold clusters of different sizes supported over various metal oxides (Al_2O_3, TiO_2, and SiO_2) were studied to different CO oxidation conditions that were investigated by Bokhoven et al. using in situ X-ray absorption spectroscopy at the Au L3 edge [51]. During catalysis, only phase detected in all catalysts was Au0 with no other oxidation states. For the most active catalyst with small size gold particles (Au/Al_2O_3), difference in the electronic structure of the gold clusters in changing reaction conditions was observed by XANES spectroscopy and attributed to the adsorption of CO on the metallic gold clusters.

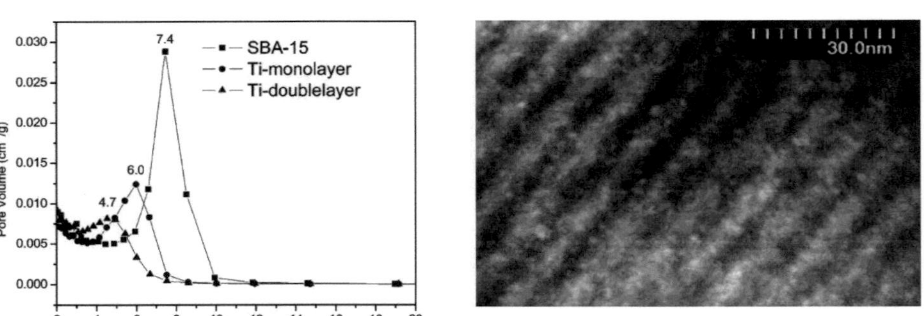

Figure 3.
Pore-size distribution as function of the number of TiO_2 layers (left), Z-contrast TEM image of ultra-small Au NPs on ordered mesoporous materials (right). The bright spots (0.8–1.0 nm) correspond to Au NPs. Reproduced with permission, copyright 2018, American Chemical Society.

Further FEFF8 calculations confirmed that the changes in the XANES signature of Au/Al_2O_3 could be due to the back donation of d-electrons into the π^* orbitals of CO. For Au/Al_2O_3, Au-O backscattering was also seen in the EXAFS, suggesting weak cluster-support interactions. Nevertheless for Au/SiO_2 and Au/TiO_2, the structure of the gold clusters remained unaffected throughout all experiments.

It is well recognized that several-supported metal particles are subjected to the geometric/electronic variations when treated in oxidizing atmosphere at high temperatures [52–53]. Recently, Huang et al. examined the effect of oxygen treatment on the catalytic performance of Au/SiO_2 catalysts toward CO oxidation [54]. The Au/SiO_2 catalysts were synthesized using deposition-precipitation practice. It was observed that, the as-prepared Au/SiO_2 catalyst executes a poor catalytic activity, which is because of the relatively large size of Au particles. After treating the catalyst in O_2 at temperatures higher than 800°C efficiently enhances the catalytic activity with the agglomeration of Au particles. However, TEM results reveal the co-existence of uniformly well-dispersed ultrafine Au particles over the surface. XPS results disclose that after the oxygen pretreatment at temperatures above 800°C, the Au 4f binding energy moves to higher binding energy. Remarkably, dealing of the Au/SiO_2 catalyst in He at 800°C also displays the similar geometric and electronic structure changes of Au particles. Therefore, the enhancement effect as that in O_2, indicates that the Au-O_2 interactions at high temperatures do not contribute greatly to the progress of catalytic activity. The authors proposed that evaporation-deposition mechanism of gold particles treated at high temperatures, which accounts for the formation of ultrafine Au particles, is responsible for the improvement of the catalytic activity (**Figure 4**).

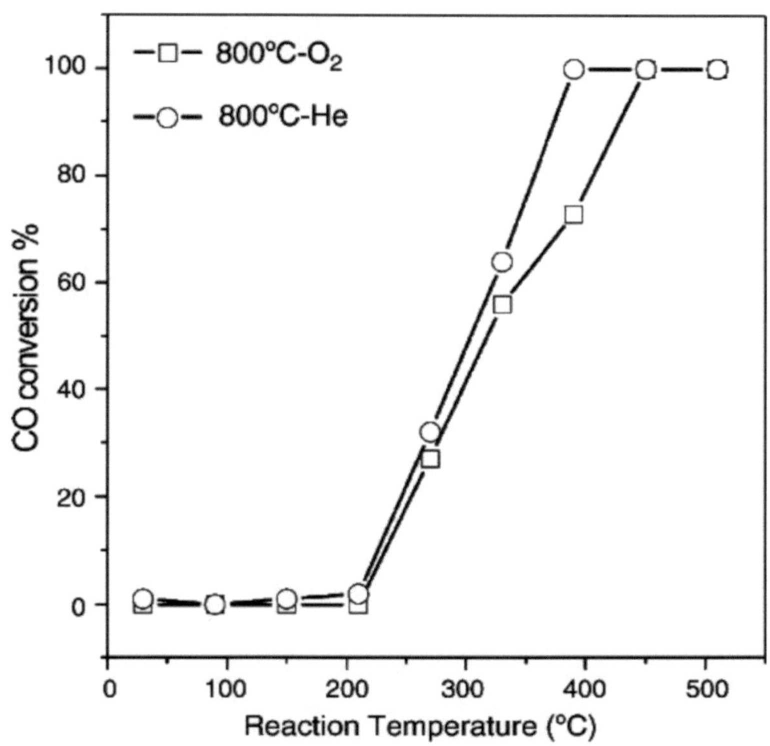

Figure 4.
The CO conversion of Au/SiO_2-O_2-800 and Au/SiO_2-He-800 catalysts. Reproduced with permission, copyright 2018, Elsevier.

MCM-41 was fused with Au NPs by Mokhonoana et al. [55] using ethylenediamine through a modified deposition-precipitation method, where ethylenediamine operated as both base and complexing agent for the Au(III) species. Synthesis involves the mixing aqueous solution of $HAuCl_4$ separately with both the as-synthesized MCM-41 (still containing CTABr) and the calcined MCM-41 for 1 h. pH was adjusted to 10 using 1 M ethylenediamine solution. Resulting slurry was agitated at room temperature for 13 h, and the recovered solids were filtered, washed with warm water to eliminate the chloride ions, dried and calcined at 500°C for 12 h (for the as-synthesized MCM-41) or 400°C for 4 h (for the calcined MCM-41).

It has been observed that, subsequent catalysts with (4 and 5% nominal Au loading, respectively), having average Au NPs of 12 and 10 nm, respectively, holds the ordered structure and high surface area of the MCM-41 material. Nevertheless, calcination at 500°C results in aggregation and migration of Au NPs to the surface. Still, both catalysts show good activity in the CO oxidation at T > 250°C.

Huang et al. investigated that, NaOH additive to Au/SiO_2 catalyst considerably improves the catalytic activity of inert in catalyzing CO oxidation at temperatures below 150°C, while $Au/NaOH/SiO_2$ catalyst with a NaOH:Au atomic ratio of 6 is active at room temperature [56]. It has been witnessed that both particle size distribution and the electronic structure of Au NPs remain equal in Au/SiO_2 and $Au/NaOH/SiO_2$ catalysts, where hydroxyls excite the activation of O_2 on "inert" Au NPs, which benefits to catalyze CO oxidation even at room temperature. Further, density functional theory (DFT) calculation results also proves the defining role of COOH in hydroxyls-induced activation of O_2 on the Au(111) surface.

The impact of pretreatments effect for gold supported over hexagonal mesoporous silica (HMS) with He, O_2, and H_2 on the physicochemical and catalytic properties has been studied by Pestryakov et al. [57]. It has been investigated that, gold supported on mesoporous silica forms different states such as Au^{3+} and Au^{+} ions, neutral, partly charged gold clusters, and metal NPs of varied sizes. It is noted that as-prepared catalyst with Au^{3+} ions does not show any catalytic activity in CO oxidation. While reduced pretreated catalysts increases the catalytic activity and its oxidative treatments deactivate the catalyst. Catalytic tests demonstrate that, reduced samples contain several regions with various catalytic behaviors at 20–200°C, 200–400°C, and >400°C; this is because of the co-existence of gold active sites of varied forms. Further, a comparative analysis using XPS, UV-visible spectroscopy, FTIR and catalytic data shows that $Au^{\delta+}{}_n$ clusters or/and Au^{+} ions are responsible for the activity in the low-temperature region (i.e. <200°C); while neutral Au_n clusters are found to be active in the temperature range 200–400°C; and Au NPs catalyze the high-temperature CO oxidation.

Vinod et al. have proposed and confirmed the theory of active sites. They have examined the role of interfaces for the CO oxidation reaction for trisoctahedral (TOH) Au NPs adorned with nano oxides of CeO_2 and TiO_2 encapsulated in porous silica system [58]. The TEM images display that size of TOH morphology of the Au NPs is ~70 nm, which shows the high index facets with unvarying defect sites. Its atomic model shows the existence of {221} and {331} high index planes, which carries {111} terrace and mono atomic {110} step atoms as reported earlier [59]. Further, line profile analysis of HRTEM image from the surface of the TOH particle reveals the existence of step-terrace geometry. These oxides adorned and silica-encapsulated system were found to display substantial activity and stability for CO oxidation at room temperature; nevertheless, the Au particle size was above the optimum range. This is due to the conservation of morphology and thereby the active centers due to encapsulation.

1.1.3 Synthesis of Au/SiO_2 using cationic gold complex $[Au(en)_2]Cl_3$ (en = ethylenediamine)

Dai et al. reported a unique deposition-precipitation (DP) method for the preparation of highly active Au catalysts supported over mesoporous silica (SBA-15) using a gold cationic complex precursor $[Au(en)_2]^{3+}$ via a wet chemical process [60] (**Figure 5**). This new DP procedure comprises the use a cationic gold precursor instead of anionic $AuCl_4^-$, which is facilitated by ion-exchange route. The subsequent mesoporous catalyst is found to be extremely active for CO oxidation reaction at room temperature and even below 0°C. Its catalytic activity is found to be much greater than that of silica-supported Au catalysts formerly prepared through solution techniques.

In addition, pH of the gold precursor solution founds to play a key role in determining the catalytic activity through the regulation of $[Au(en)_2]^{3+}$ deprotonation reaction and the surface interaction of silica with the gold precursor (**Figure 6**). It is also observed that these mesoporous gold silica catalyst are highly resistant toward sintering because of the stabilization of Au NPs within mesopores. The authors projected that this synthesis strategy of silica-supported gold catalysts is entirely solution-based and can be applied to prepare gold catalysts supported over different kinds of silica materials (e.g., silica particles and microporous zeolites).

Gies et al. and coworkers have deposited Au NPs around 3 nm particles inside the channels of mesoporous silica-TiO_2-MCM-48 using deposition techniques [61]. It

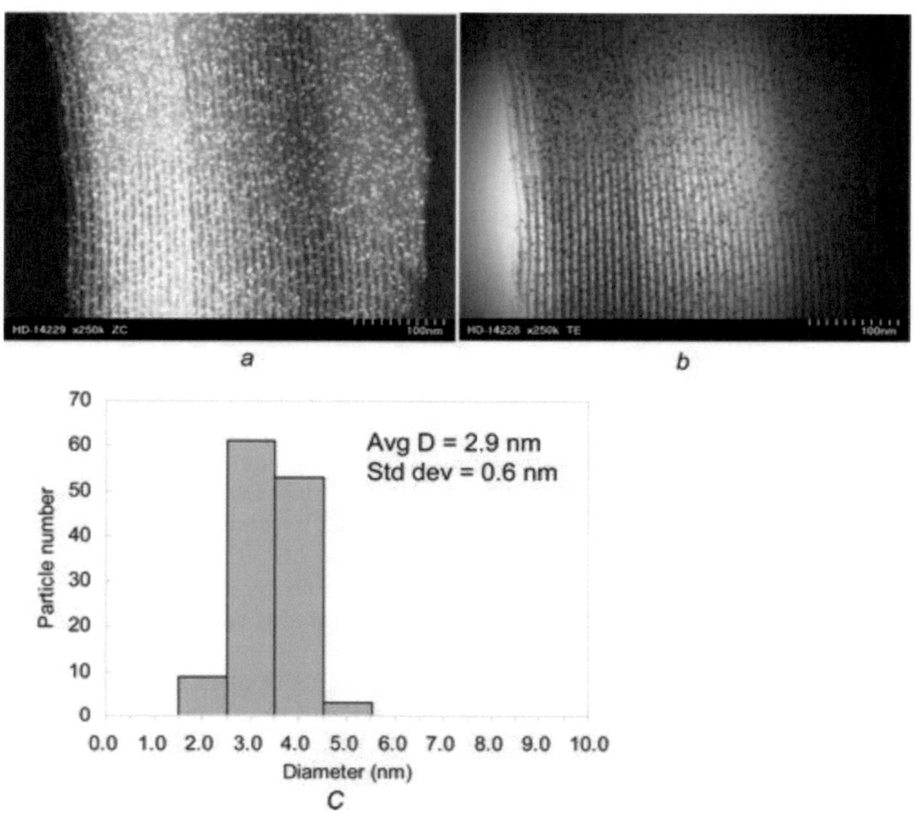

Figure 5.
TEM images [(a) dark field, (b) bright field and (c) size distribution histogram] of the Au catalyst supported on SBA-15 (synthesized at pH of 9.6 and reduced at 150 °C).

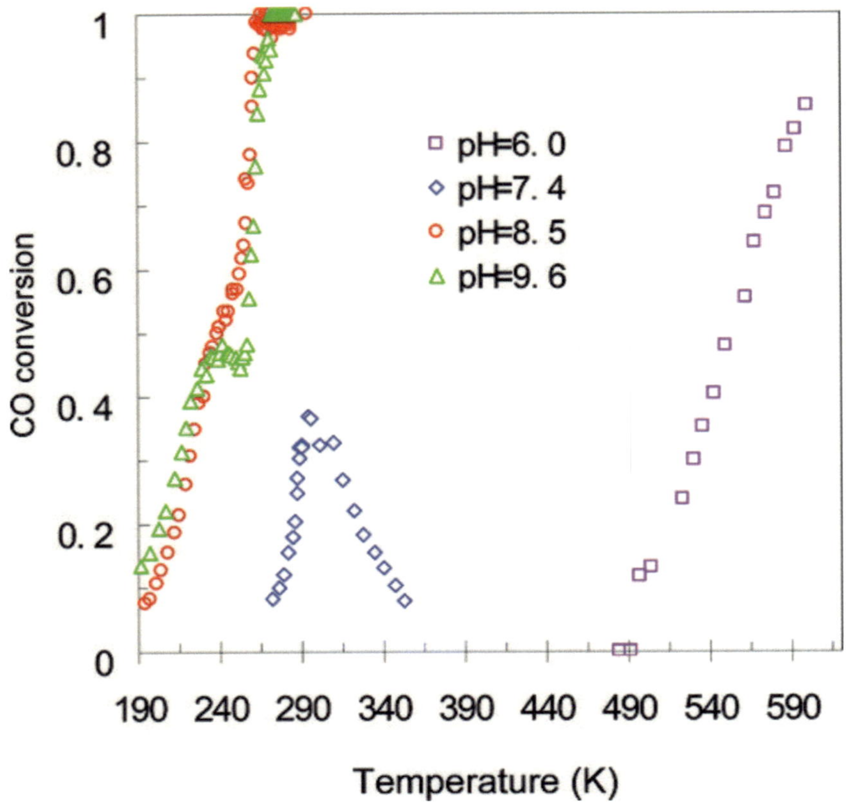

Figure 6.
Light-off curves of the Au catalysts supported on SBA-15, synthesized in the solutions of $Au(en)_2Cl_3$ with different pH values. Reproduced with permission, copyright 2018, American Chemical Society.

has been revealed that, Au NPs catalyst over mesoporous silica not only converts CO to CO_2 at 50% level at −20°C but also stable against sintering up to at least 200°C.

Dai et al. [62] described a novel method for the synthesis, characterization, and catalytic behavior of small and well-dispersed Au NPs on Au/Cab-O-Sil fumed SiO_2 and Au/MOx/SiO_2 catalysts using $Au(en)_2Cl_3$ (en = ethylenediamine) as the precursor. It has been found that, these Au/SiO_2 catalysts are extremely active for CO oxidation below 0°C. Pretreating of as-synthesized Au/SiO_2 in H_2-Ar at 150°C and in O_2-He at 500°C is found to be very cooperative for high activity with optimum gold loading of 1.1 and 2.5 wt%. Furthermore, the post-treatment of calcined (and activated) Au/SiO_2 in different media motivates the activity in CO oxidation. Moreover, the addition of metal oxide dopants also has been used to tune the catalytic activity.

Wu et al. investigated the nature of Au species over Au/SiO_2 catalyst after oxidative and reductive pretreatments and also their role in room temperature CO oxidation using operando diffuse reflectance infrared spectroscopy (DRIFT) coupled with quadruple mass spectrometry (QMS) [63]. It has been observed that, the oxidative pretreatment of catalyst leads to a cationic Au species, which is inactive for CO oxidation at rt. However, in situ reduction of the cationic Au during CO oxidation leads to the formation of Au(0) species, which is active for CO oxidation. It is recognized that the reductive pretreatment results in a Au(0) species, which have sturdier interaction with the support and thus are more active for CO oxidation than those on oxidatively treated catalyst. It is also shown that, water has two positive effects in CO oxidation reaction on Au/SiO_2: activating O_2 species and supporting the reduction of Au species.

1.1.4 Chemical vapor deposition

To elucidate the effect of metal oxide support on the catalytic activity of gold for CO oxidation, Okumura et al. have deposited gold on SiO_2, Al_2O_3, and TiO_2 with high dispersion by chemical vapor deposition (CVD) of an organo-gold complex [64]. The results show that orders of TOF values for CO oxidation at 0°C are similar among Au/Al_2O_3, Au/SiO_2, and Au/TiO_2. Further, it was demonstrated that, the deposition of gold particles on the support with strong interaction is a major key controlling factor for the evolution of catalytic activity for CO oxidation at temperature 0°C (and not at −70°C) and the nature of the support is not a dominant factor.

Similarly, silica-supported Au NPs of size 1.4 nm were prepared by organometallic chemical vapor deposition method (Au/SiO_2-CVD) by Claus et al. [65]. Furthermore, the synthesized catalysts show a notable activity for CO oxidation at low temperature.

Okumura et al. have deposited gold on Al_2O_3, SiO_2, MCM-41, TiO_2, SiO_2-Al_2O_3, and active carbon (AC) support by gas-phase grafting (GG) of an organo-gold complex with high dispersion to display the effect of support in CO oxidation [66]. It has been found that, order of TOF values for CO oxidation at 0 °C are similar among Au/Al_2O_3, Au/SiO_2, and Au/TiO_2, this clearly shows the deposition of Au NPs on the supports with strong interaction which play important role in catalytic activity. Whereas semi-conductive or reducible nature of the support is not a presiding factor for CO oxidation at 0 °C. Au/SiO_2-Al_2O_3 and Au/AC shows lower catalytic activities due to acidic and nonmetal-oxide supports respectively.

Veith et al. reported a new way to prepare small size Au NPs (2.5 nm) over a fumed silica support, using the physical vapor deposition technique of magnetron sputtering [67]. These Au/SiO_2 catalysts are found to be structurally stable when heated in air to 500°C for several weeks or during a CO oxidation reaction. However, under these annealing conditions, traditional Au/TiO_2 catalysts rapidly sinter to form large 13.9 nm gold clusters, resulting in a fivefold decrease in activity. The authors witnessed that the stability of Au/SiO_2 is usually accredited to the absence of residual impurities (ensured by the halide-free production method) and a strong bond between gold and defects at the silica surface (about 3 eV per bond) is calculated from density functional theory (DFT) calculations. Properties that make the material worthy of study include the ability to easily reactivate the catalyst, thermal stability, and the unique gold-support interactions.

1.1.5 Synthesis by dispersion of gold colloids or pre-synthesized AuNPs

McFarland et al. examined the reactivity of gold clusters (8–22 nm diameter) supported on different metal oxides TiO_2, ZnO, ZrO_2, and SiO_2 in a continuous flow reactor [68]. Synthesis involves the encapsulation of gold clusters within diblock copolymer [polystyrene$_{81,000}$-block-poly(2-vinylpyridine)$_{14,200}$] in toluene solution, impregnated onto the bulk supports, and reduced by calcination at temperature 300°C. TiO_2 > ZrO_2 > ZnO order of support-dependent sintering was observed in air at 300°C. Au nanoclusters on TiO_2 found to exhibit the highest activity for CO oxidation compared to other metal oxide supports.

1.1.6 New methods

Guczi et al. [69] have reported a new method for the preparation of Au NPs sizes of about 50–100 × 20–30 × 2–7 nm along with spheres of 5–10 nm diameter over SiO_2/Si(100) by Ar^+ ion implantation of a bulk-like Au/SiO_2/Si(100) thin film of 10 nm thickness. Photoemission spectra show that during size reduction, Au 5d valence band of Au NPs shows a change in the valence band density state, and hence the catalytic activity in the CO oxidation increases. Moreover, it has been observed that, at high-temperature catalytic tests, both size and valence band of the Au NPs get returned to its typical bulk gold, which consequently curbed its catalytic activity.

Margitfalvi et al. reported a novel process for the preparation of silica-supported nanosized gold catalysts using $HAuCl_4$ gold precursor and ammonia solution. This gold catalyst showed relatively high catalytic activity in CO oxidation in a broad temperature region [70]. It was identified that, uniform well deposition of gold over silica support takes place because of the electrostatic interaction between the silica surface and the positively charged gold amine complexes resulting due to the being of ammonia solution, which avoids the growth of particle during hydrogen treatment at 350°C; this subsequently results in a highly dispersed gold particles over the silica surface. Margitfalvi et al. also observed that the presence of moisture in reactant mixture increases the catalytic activity of Au/SiO_2 catalysts, while TOS study reveals the unstability of Au/SiO_2 catalysts.

Coperet et al. and coworkers, described a novel route for the direct preparation of Au NP (1.8 nm) on partially dehydroxylated silica at 700°C [$SiO_{2\text{-}(700)}$] by surface organometallic chemistry [71]. Method involves the controlled formation of well-defined and dispersed Au NPs functionalized over passivated silica surface with a AuI complex, $\{Au[N(SiMe_3)_2]\}_4$, followed by mild reduction under H_2 at 300°C (**Figure 7**). This methodology results in a very efficient tailor-made gold

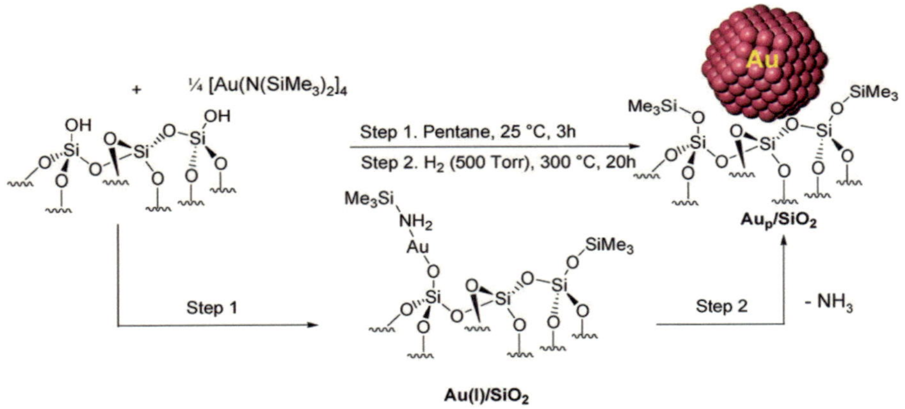

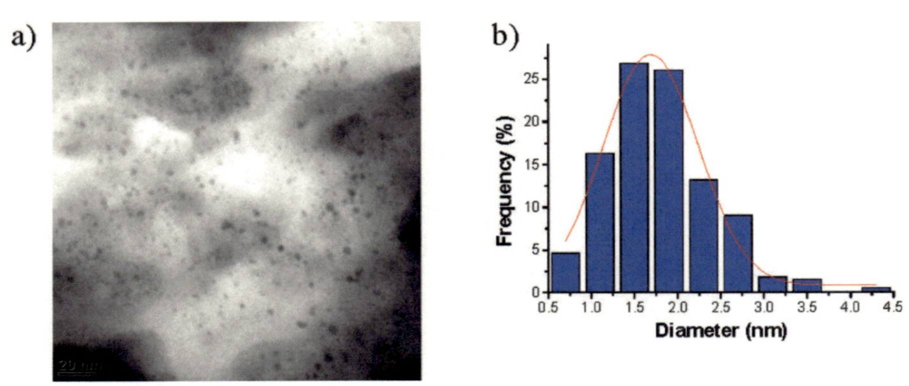

Figure 7.
(A) Schematic route for the preparation of AuNP on passivated silica by SOMC (2.79 wt% Au). (B) (a) HR-TEM image and (b) size distribution histogram of AuNPs over passivated silica by SOMC.

DOI: http://dx.doi.org/10.5772/intechopen.80620

catalyst, suitable not only for liquid phase aerobic epoxidation of trans-stilbene but also for oxidation of CO.

Recently, we have described a facile route for the synthesis of Au NPs of different average sizes (ranging from 1 to 2, 3 to 5, and 11 to 13 nm) on to the fibrous silica nanospheres (KCC-1) either by immobilizing pre-made Au NPs on amine-functionalized KCC-1 (Au/KCC-1-NH_2) or by grafting $HAuCl_4$ on Au/KCC-1-NH_2 followed by reduction by $NaBH_4$, CO, and citrate [72]. It's been important to note that size and the location of the Au NPs over the support were found to influence by the preparation method (**Figure 8**). These catalysts were then tested for the oxidation of CO. The catalytic activity of the Au NPs over KCC-1 was proved to be size dependent (**Figure 9**).

Recently, Fan et al. have prepared an exceptional pollen-structured hierarchically meso-/macroporous silica spheres (PHMSs) created from unique rape pollen grains and a triblock copolymer poly(ethylene oxide-block-propylene oxide-block-ethylene oxide) (P123) as templates [73]. The PHMSs are enclosed of mesoporous walls, which create a macroporous structure repeating the net-like morphology of the pollen grain exine. This newly structured catalyst found to exhibit a high BET surface area of 626 m^2 g^{-1} and uniform mesopores of average pore size less than 5 nm. Au NPs supported on PHMSs (Au 6.8 wt%) show a higher CO conversion at an ambient temperature (30°C), compared to those supported on typical mesoporous silica SBA-15. Furthermore, the effects of silica support on the catalytic activity have been discovered from the calculation of diffusion coefficients and simulation of gas diffusion in both PHMSs and SBA-15. Catalytic performance for CO oxidation proves that the hierarchical PHMSs facilitate better gas diffusion than the nonhierarchical SBA-15.

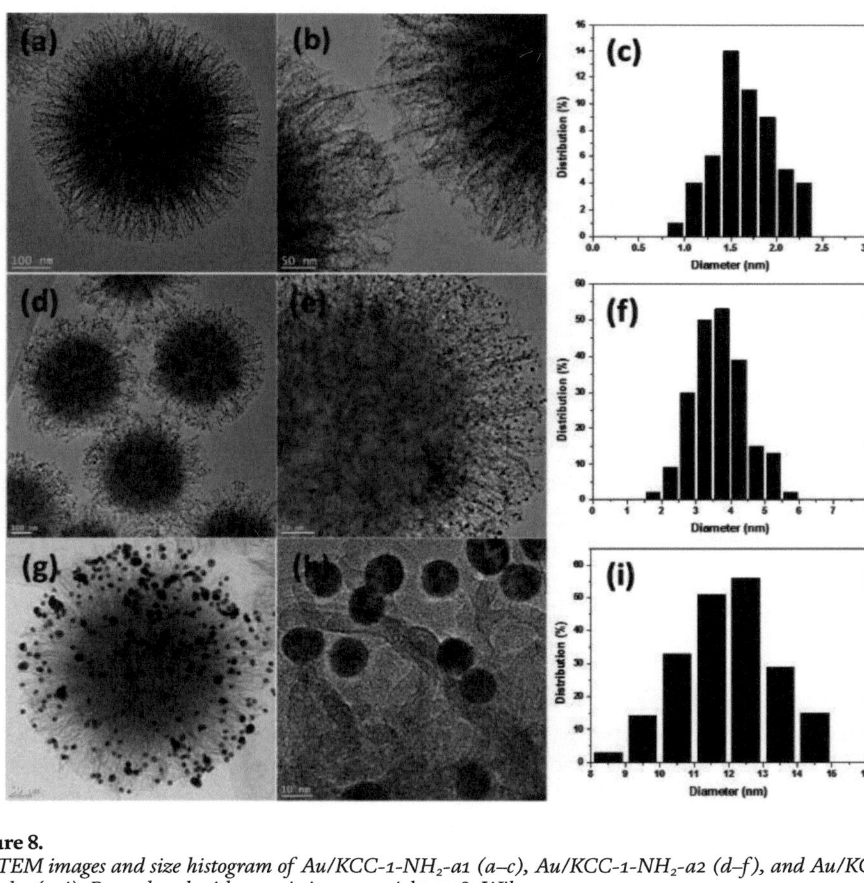

Figure 8.
HR-TEM images and size histogram of Au/KCC-1-NH_2-a1 (a–c), Au/KCC-1-NH_2-a2 (d–f), and Au/KCC-1-NH_2-b1 (g–i). Reproduced with permission, copyright 2018, Wiley.

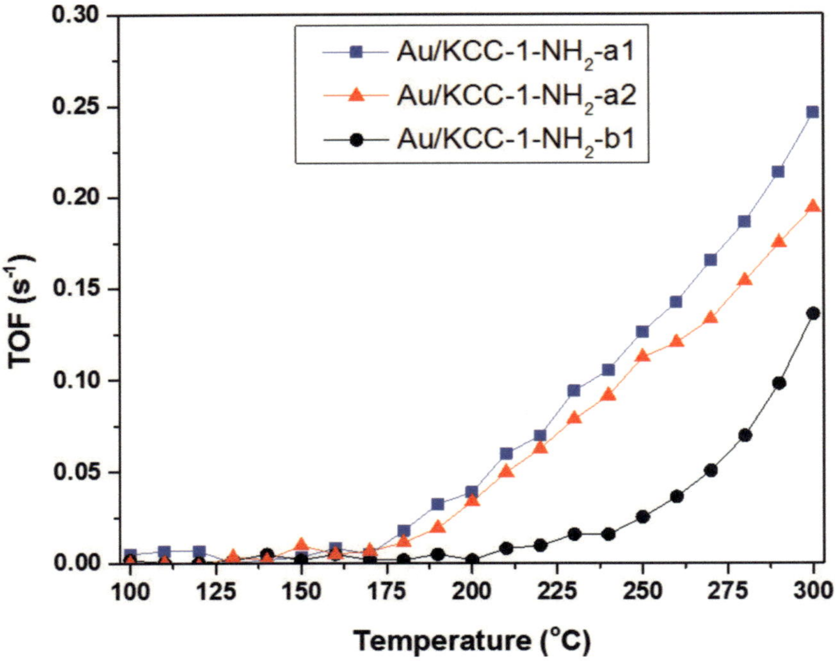

Figure 9.
Turn over frequency (TOF S^{-1}) for CO oxidation over Au/KCC-1-NH_2-a1, Au/KCC-1-NH_2-a2 and Au/KCC-1-NH_2-b1 after H_2 treatment at 300°C for 1 h. Reproduced with permission, copyright 2018, Wiley.

Yang et al. reported an improved strategy for the preparation of SBA-15-supported metal nanoparticles [74]. Supported Au nanoparticle catalyst Au/SBA-15 exhibited significantly higher activities in the oxidation of CO by molecular oxygen after adding polyvinylpyrrolidone (PVP) to the reaction system. Earlier, it has been reported that, PVP could strongly interact with the soluble Au NPs in water solvent, leading to high activity in oxidation reaction [75–76].

Recently, Behm et al. have developed an improved catalytic system for the stability of metal oxide-supported Au catalysts for low-temperature CO oxidation reaction [77]. Effect of mesoporous silica SBA-15 substrate on the catalytic activity and stability of Au/TiO_2 catalysts has been explored, where the catalyst system composed of a SBA-15 support surface modified by a monolayer of TiOx and Au NPs on top. Later, TiOx surface layers were systematically increased without changing the Au loading and particle size. Kinetic measurements were calculated at three different temperatures (30, 80, and 180°C). It was noted that catalyst surface modification with Ti has distinct effect, the performance of these catalysts increases significantly with Ti concentration and also with reaction temperature.

1.2 Au-M bimetallic nanoparticles on various silica support for the CO oxidation reaction

Recent research reveals that bimetallic nanoparticles often exhibit better catalytic performances compared with the monometallic counterparts. This has led to a great interest in both academic and industrial fields. The interaction between the two components of bimetallic catalysts is mostly understood in terms of assemble and ligand effects, where one component may act as a spacer to separate the active sites [78] or as an electronic modifier to the other component [79]. The catalytic performances of bimetallic nanoparticles are subject to depend on several factors, such as nanostructure, surface composition, particle size, and shape. The most important features which differentiate a bimetallic catalyst from monometallic catalysts are the surface compositions and tunable nanostructures. A bimetallic nanostructure could be a random alloy, a core-shell structure, or just mixed monometallic nanoparticles, which give rise to different catalytic activities.

There are several literature reports in account, for the effect of second metals like Ti [78–88], Cu [89–92], Fe [93–95], Ag [96–98], Ce [99–100], Al [101–102], Co [103], Pd [104], Pt [105], and Sn [106] to the Au NPs for the activity of CO oxidation. It has been established that, synergistic effects of both the metals improve the catalytic performance in CO oxidation reaction. Type of bimetallic catalysts, support, size of the Au NPs, and the gold loading are showing remarkable effects on the activity of the reaction (**Table 2**).

Bimetallic catalyst	Support	Au loading (wt%)	Au size (nm)	Temperature (°C)	Ref.
Au-Ti	Silica aerogel	0.1–20	2.0	−50–70	[80]
	Silica aerogel	3.4–6.7	1.4–6.4	−50–70	[81]
	Silica aerogel	5.0	2.0		[82]
	SBA-15	1.7–3.6	3.0–15	200	[83]
	SBA-15	—	0.8–1.0	−40 (T_{50})	[84]
	SBA-15	1	3–4	50	[85]
	MCM-48	3.0–7.5	<1	17–27	[86]
	MCM-48	—	—	<0	[87]
	Silica	1.4–2.0	6.6–7.2	160	[88]
	MCM-41	—	<3	350	[89]
	Silica SBA-15	1.8–2.5	2.9–10.2	120 (T_{50})	[90]
Au-Cu	SiO_2	2 (Au-Cu)	14 (Au-Cu colloidal)	150	[91]
	SiO_2	2 (Au-Cu)	10	300	[92]
	SiO_2	4.6 (Au-Cu)	4 (Au-Cu)	25–300	[93]
	SBA-15	6 (Au-Cu)	~3	0	[94]
Au-Fe	SBA-15	0.5–0.8	<4	20	[95]
	Hexagonal mesoporous silica (HMS)	0.4–5.1	3.3–4.5	25 (Au 5.1 wt%)–300 (Au 0.4 wt%)	[96]
	Mesoporous silica	0.4–2.0	>5	−5	[97]
Au-Ag	Silica	4.9–5.6 (Au-Ag)	~3 (Au-Ag)	0	[98]
	MCM	8 (Au-Ag)	6–7	RT	[99]
	MCM	5 (Au-Ag)	4–6 (Au-Ag)	30	[100]
Au-Ce	SBA-15	1.3–2.7	4.4–26	160–170	[101]
	SBA-15	2	3–6	150–350	[102]
Au-Al	SBA-15	3–17	3	80	[103]
	SBA-15	15–20	2.7	80	[104]

Bimetallic catalyst	Support	Au loading (wt%)	Au size (nm)	Temperature (°C)	Ref.
Au-Co	SBA-15	2	<5	0	[105]
Au-Pd	Silica fume	5.9–77.3	4–12	150–250	[106]
Au-Pt	Silica	—	3.5–4.6	280	[107]
Au-Sn	SiO_2	2.6–3.0	—	200	[108]
Au-Co Au-Fe	MCM-41 Aerosol-silica	5	3–8	>400	[109]
Au-Ti Au-Ce	SBA-15	0.6–2.5	2.9–5.0	Au-Ti 150 Au-Ce 50	[110]
Au-Co Au-Fe Au-Cu	Stober silica	1	5 (Au-M)	300	[111]
Au-Fe Au-Ce Au-Ti	Hexagonal mesoporous silica (HMS)	2.8–3.4	0.5–6	Au-Fe 150 Au-Ce 200 Au-Ti 200	[112]

Table 2.
Au-M bimetallic catalysts with different types of silica supports for the CO oxidation reaction.

2. Conclusions

The synthesis of stable Au NPs and metal oxide support introduces new possibilities for catalysis since their properties are neither those of the bulk materials or of the isolated atoms. Optimum activity for, in particular, CO oxidation is practical for gold clusters in the 2–5 nm range. Furthermore, these gold nanoparticles can have a strong interaction with the support and an electron relocation may happen within either directions. These interactions can result in the formation of stabilized surface charged species, which act as Lewis acids comparable to as of salts in solution or in the corresponding transition metal complexes.

Mono- and bimetallic gold nanocatalysts have been made over a variety of metal oxides supports. In particular, silica is being considered a suitable support for Au NPs using appropriate synthesis methods. Au NPs can be supported over a series of different mesoporous silicas, such as SBA-15, SBA-16, MCM-41, and MCM-48. These silica-supported gold nanocatalysts are found to exhibit exceptional catalytic properties. Most common methods for the modification of silica support is either by direct or post-grafting functionalization with amine or thiol groups, where these groups show dual performances, as ligand for improving the interaction between the metal precursors and the silica surface; and mild reductants under basic conditions for in situ reduction of Au(III).

Several active silica-supported gold nanocatalysts have been prepared by functionalizing the mesoporous silica support by direct or subsequent incorporation of the gold precursor. Major problems related to these processes include particle size, gold loading, uniform dispersion Au NPs, and ligand removal during calcination, which contribute to the aggregation of Au NPs, metal leaching, and low stability.

Finally, it is precise to say that silica-supported gold nanocatalyst has become a hot topic of research for CO oxidation; conversely, there are still many challenges ahead for the improvement of silica-supported gold nanocatalyst filling the main necessities of any catalyst such as easy and low-cost synthesis method, high activity, selectivity, and greater stability at lower temperature. Furthermore, the proof of identity for the active gold species is still a challenging job for CO oxidation reactions catalyzed by gold.

Acknowledgements

The authors acknowledge the support from the Ministry of Higher Education, Saudi Arabia, in the establishment of the Center of Research Excellence in Petroleum Refining & Petrochemicals at King Fahd University of Petroleum & Minerals (KFUPM).

Conflict of interest

The authors declare that the research was conducted in the absence of any commercial or financial relationship that could be construed as a potential conflict of interest.

Author details

Ziyauddin S. Qureshi* and EA Jaseer
Center for Refining and Petrochemicals, King Fahd University of Petroleum and Minerals (KFUPM), Dhahran, Saudi Arabia

*Address all correspondence to: zqureshi@kfupm.edu.sa

References

[1] Hutchings GJ. Vapor phase hydrochlorination of acetylene: Correlation of catalytic activity of supported metal chloride catalysts. Journal of Catalysis. 1985;**96**:292-295

[2] Haruta M, Kobayashi T, Sano H, Yamada N. Novel gold catalysts for the oxidation of carbon-monoxide at a temperature far below 0°C. Chemistry Letters. 1987;**16**:405-408

[3] Pina DC, Falletta E, Prati L, Rossi M. Selective oxidation using gold. Chemical Society Reviews. 2008;**37**:2077-2095

[4] Gutiérrez LF, Hamoudi S, Belkacemi K. Selective production of lactobionic acid by aerobic oxidation of lactose over gold crystallites supported on mesoporous silica. Applied Catalysis A: General. 2011;**402**:94-103

[5] Claus P. Heterogeneously catalysed hydrogenation using gold catalysts. Applied Catalysis A: General. 2005;**291**:222-229

[6] Corma A, Serna P. Chemoselective hydrogenation of nitro compounds with supported gold catalysts. Science. 2006;**313**:332-334

[7] Fu Q, Weber A, Flytzani-Stephanopoulos M. Nanostructured Au-CeO_2 catalysts for low-temperature water-gas shift. Catalysis Letters. 2001;**77**:87-95

[8] Idakiev V, Tabakova T, Tenchev K, Yuan ZY, Ren TZ, Vantomme A, et al. Gold nanoparticles supported on ceria-modified mesoporous-macroporous binary metal oxides as highly active catalysts for low-temperature water-gas shift reaction. Journal of Materials Science. 2009;**44**:6637-6643

[9] Flytzani-Stephanopoulos M, Fu Q, Saltsburg H. Active nonmetallic Au and Pt species on ceria-based water-gas shift catalysts. Science. 2003;**301**:935-938

[10] Nkosi B, Coville NJ, Hutchings GJ, Adams MD, Friedl J, Wagner FE. Hydrochlorination of acetylene using gold catalysts–A study of catalyst deactivation. Journal of Catalysis. 1991;**128**:366-377

[11] Conte M, Carley AF, Heirene C, Willock DJ, Johnston P, Herzing AA, et al. Hydrochlorination of acetylene using a supported gold catalyst: A study of the reaction mechanism. Journal of Catalysis. 2007;**250**:231-239

[12] Edwards JK, Solsona BE, Landon P, Carley AF, Herzing A, Kiely CJ, et al. Direct synthesis of hydrogen peroxide from H_2 and O_2 using TiO_2-supported Au-Pd catalysts. Journal of Catalysis. 2005;**236**:69-79

[13] Ueda A, Oshima T, Haruta M. Reduction of nitrogen monoxide with propene in the presence of oxygen and moisture over gold supported on metal oxides. Applied Catalysis B: Environmental. 1997;**12**:81-93

[14] Nguyen LQ, Salim C, Hinode H. Roles of nano-sized au in the reduction of NOx by propene over Au/TiO_2: An in situ drifts study. Applied Catalysis B: Environmental. 2010;**96**:299-306

[15] Teles JH, Brode S, Chabanas M. Cationic gold(I) complexes: Highly efficient catalysts for the addition of alcohols to alkynes. Angewandte Chemie International Edition. 1998;**37**:1415-1418

[16] Bond GC, Thompson D. Gold-catalysed oxidation of carbon monoxide. Gold Bulletin. 2000;**33**:41-50

[17] Park ED, Lee D, Lee HC. Recent progress in selective CO removal in

a H_2-rich stream. Catalysis Today. 2009;**139**:280-290

[18] Physical constants of inorganic compounds. In: Handbook of Chemistry and Physics. 87th ed. Boca Raton: CRC Press, Taylor & Francis Group; ChemCatChem. 2011;**3**:24-65

[19] EarthTrends of the World Resources Institute. ChemCatChem 2011;**3**:24-65. Available from: http://earthtrends.wri.org/searchable_db/index.php?theme=3

[20] Haruta M. Size- and support-dependency in the catalysis of gold. Catalysis Today. 1997;**36**:153-166

[21] Li L, Gao Y, Li H, Zhao Y, Pei Y, Chen Z, et al. CO oxidation on TiO_2 (110) supported subnanometer gold clusters: Size and shape effects. Journal of the American Chemical Society. 2013;**135**:19336-19346

[22] Comotti M, Li WC, Spliethoff B, Schìth F. Support effect in high activity gold catalysts for CO oxidation. Journal of the American Chemical Society. 2006;**128**:917-924

[23] Kireev AS, Mukhin VM, Kireev SG, Klushin VN, Tkachenko SN. Preparation and properties of modified hopcalite. Russian Journal of Applied Chemistry. 2009;**82**:169-171

[24] Ohyama J, Koketsu T, Yamamoto Y, Arai S, Satsuma A. Preparation of TiO_2-supported twinned gold nanoparticles by CO treatment and their CO oxidation activity. Chemical Communications. 2015;**51**:15823-15826

[25] Xie X, Li Y, Liu ZQ, Haruta M, Shen W. Low-temperature oxidation of CO catalysed by Co_3O_4 nanorods. Nature. 2009;**458**:746-749

[26] Qi J, Chen J, Li G, Li S, Gao Y, Tang Z. Facile synthesis of core–shell Au@ CeO_2 nanocomposites with remarkably enhanced catalytic activity for CO oxidation. Energy & Environmental Science. 2012;**5**:8937-8941

[27] Liu Y, Jia CJ, Yamasaki J, Terasaki O, Schìth F. Highly active iron oxide supported gold catalysts for CO oxidation: How small must the gold nanoparticles be? Angewandte Chemie International Edition. 2010;**49**:5771-5775

[28] Kimura T, Kuroda K. Mesoporous silica. In: Reedijk J, Poeppelmeier K, editors. Comprehensive Inorganic Chemistry II. 2nd ed. Amsterdam: Elsevier; 2013. pp. 133-150

[29] Corma A. Microporous to mesoporous molecular sieve materials and their use in catalysis. Chemical Reviews. 1997;**97**:2373-2419

[30] Gutirrez LF, Hamoudi S, Belkacemi K. Synthesis of gold catalysts supported on mesoporous silica materials: Recent developments. Catalysts. 2011;**1**:97-154

[31] Haruta M. Catalysis of gold nanoparticles deposited on metal oxides. CATTECH. 2002;**6**:102-115

[32] Daniel MC, Astruc D. Gold nanoparticles: Assembly, supramolecular chemistry, quantum-size-related properties, and applications toward biology, catalysis, and nanotechnology. Chemical Reviews. 2004;**104**:293-346

[33] Gates BC, Fierro-Gonzalez JC. Catalysis by gold dispersed on supports: The importance of cationic gold. Chemical Society Reviews. 2008;**37**:2127-2134

[34] Corma A, Garcia H. Supported gold nanoparticles as catalysts for organic reactions. Chemical Society Reviews. 2008;**37**:2096-2126

[35] Pyykko P. Theoretical chemistry of gold. III. Chemical Society Reviews. 2008;**37**:1967-1997

[36] Haruta M, Date M. Advances in the catalysis of Au nanoparticles. Applied Catalysis A: General. 2001;**222**:427-437

[37] Hashmi ASK, Hutchings GJ. Gold catalysis. Angewandte Chemie International Edition. 2006;**45**:7896-7936

[38] Yang C, Kalwei M, Schuth F, Chao K. Gold nanoparticles in SBA-15 showing catalytic activity in CO oxidation. Applied Catalysis A: General. 2003;**254**(2):289-296

[39] Bore MT, Pham HN, Ward TL, Datye AK. Role of pore curvature on the thermal stability of gold nanoparticles in mesoporous silica. Chemical Communications. 2004:2620-2621

[40] Bore MT, Pham HN, Switzer EE, Ward TL, Fukuoka A, Datye AK. The role of pore size and structure on the thermal stability of gold nanoparticles within mesoporous silica. Journal of Physical Chemistry B. 2005;**109**:2873-2880

[41] Chi YS, Lin HP, Mou CY. CO oxidation over gold nanocatalyst confined in mesoporous silica. Applied Catalysis A: General. 2005;**284**:199-206

[42] Rombi E, Cutrufello MG, Cannas C, Casu M, Gazzoli D, Occhiuzzi M, et al. Modifications induced by pretreatments on Au/SBA-15 and their influence on the catalytic activity for low temperature CO oxidation. Physical Chemistry Chemical Physics. 2009;**11**:593-602

[43] Cutrufello MG, Rombi E, Cannas C, Casu M, Virga A, Fiorilli S, et al. Synthesis, characterization and catalytic activity of Au supported on functionalized SBA-15 for low temperature CO oxidation. Journal of Materials Science. 2009;**44**:6644-6653

[44] Rombi E, Cutrufello MG, Monaci R, Cannas C, Gazzoli D, Onida B, et al. Gold nanoparticles supported on conventional silica as catalysts for the low-temperature CO oxidation. Journal of Molecular Catalysis A: Chemical. 2015;**404-405**:83-91

[45] Lin SD, Bollinger M, Vannice MA. Low temperature CO oxidation over Au/TiO_2 and Au/SiO_2 catalysts. Catalysis Letters. 1993;**17**:245-262

[46] Overbury SH, Ortiz-Soto L, Zhu H, Lee B, Amiridis MD, Dai S. Comparison of Au catalysts supported on mesoporous titania and silica: Investigation of Au particle size effects and metal-support interactions. Catalysis Letters. 2004;**95**:99-106

[47] Date M, Okumura M, Tsubota S, Haruta M. Vital role of moisture in the catalytic activity of supported gold nanoparticles. Angewandte Chemie, International Edition. 2004;**43**(16):2129-2132

[48] Yan W, Chen B, Mahurin SM, Hagaman EW, Dai S, Overbury SH. Surface sol-gel modification of mesoporous silica materials with TiO_2 for the assembly of ultrasmall gold nanoparticles. Journal of Physical Chemistry B. 2004;**108**:2793-2796

[49] Ichinose I, Senzu H, Kunitake T. Stepwise adsorption of metal alkoxides on hydrolyzed surfaces: A surface sol-gel process. Chemistry Letters. 1996;**25**:831-832

[50] Budroni G, Corma A. Gold-organic-inorganic high-surface-area materials as precursors of highly active catalysts. Angewandte Chemie, International Edition. 2006;**45**(20):3328-3331

[51] Weiher N, Bus E, Delannoy L, Louis C, Ramaker DE, Miller JT, et al. Structure and oxidation state of gold on different supports under various CO oxidation conditions. Journal of Catalysis. 2006;**240**:100-107

[52] Stulga JE, Wynblatt P, Tien JK. Particle splitting and redispersion phenomena in model alumina-supported platinum catalysts. Journal of Catalysis. 1980;**62**:59-69

[53] Wang T, Schmidt LD. Morphology and redispersion of Ir on SiO_2 in oxidizing and reducing atmospheres. Journal of Catalysis. 1980;**66**:301-315

[54] Qian K, Jiang Z, Huang W. Effect of oxygen treatment on the catalytic activity of Au/SiO_2 catalysts. Journal of Molecular Catalysis A: Chemical. 2007;**264**:26-32

[55] Mokhonoana MP, Coville NJ, Datye AK. Small Au nanoparticles supported on MCM-41 containing a surfactant. Catalysis Letters. 2010;**135**:1-9

[56] Qian K, Zhang W, Sun H, Fang J, He B, Ma Y, et al. Hydroxyls-induced oxygen activation on "inert" Au nanoparticles for low-temperature CO oxidation. Journal of Catalysis. 2011;**277**:95-103

[57] Bogdanchikova N, Pestryakov A, Tuzovskaya I, Zepeda TA, Farias MH, Tiznado H, et al. Effect of redox treatments on activation and deactivation of gold nanospecies supported on mesoporous silica in CO oxidation. Fuel. 2013;**110**:40-47

[58] Sreedhala S, Maheshwari S, Betsy KJ, Vinod CP. Large trisoctahedral Au nanoparticles encapsulated inside porous silica catalyses CO oxidation at room temperature: Probing the effect of encapsulation and the role of step atoms and interfaces. Applied Catalysis A: General. 2016;**524**:1-7

[59] Horváth A, Beck A, Stefler G, Benkó T, Sáfrán G, Varga Z, et al. Silica-supported Au nanoparticles decorated by CeO_2: Formation, morphology, and CO oxidation activity. Journal of Physical Chemistry C. 2011;**115**:20388-20398

[60] Zhu HG, Liang CD, Yan WF, Overbury SH, Dai S. Preparation of highly active silica-supported Au catalysts for CO oxidation by a solution-based technique. Journal of Physical Chemistry B. 2006;**110**:10842-10848

[61] Bandyopadhyay M, Korsak O, van den Berg MWE, Grunert W, Birkner A, Li W, et al. Gold nano-particles stabilized in mesoporous MCM-48 as active CO-oxidation catalyst. Microporous and Mesoporous Materials. 2006;**89**:158-163

[62] Zhu H, Ma Z, Clark JC, Pan Z, Overbury SH, Dai S. Low-temperature CO oxidation on Au/fumed SiO_2-based catalysts prepared from $Au(en)_2Cl_3$ precursor. Applied Catalysis A: General. 2007;**326**:89-99

[63] Wu ZL, Zhou SH, Zhu HG, Dai S, Overbury SH. Drifts-QMS study of room temperature CO oxidation on Au/SiO_2 catalyst: Nature and role of different Au species. Journal of Physical Chemistry C. 2009;**113**:3726-3734

[64] Okumura M, Nakamura S, Tsubota S, Nakamura T, Azuma M, Haruta M. Chemical vapor deposition of gold on Al_2O_3, SiO_2, and TiO_2 for the oxidation of CO and of H_2. Catalysis Letters. 1998;**51**:53-58

[65] Schimpf S, Lucas M, Mohr C, Rodemerck U, Bruckner A, Radnik J, et al. Supported gold nanoparticles: In-depth catalyst characterization and application in hydrogenation and oxidation reactions. Catalysis Today. 2002;**72**:63-78

[66] Okumura M, Tsubota S, Haruta M. Preparation of supported gold catalysts by gas-phase grafting of gold acethylacetonate for low-temperature oxidation of CO and of H_2. Journal of Molecular Catalysis A: Chemical. 2003;**199**:73-84

[67] Veith GM, Lupini AR, Rashkeev S, Pennycook SJ, Mullins DR, Schwartz V, et al. Thermal stability and catalytic activity of gold nanoparticles supported on silica. Journal of Catalysis. 2009;**262**:92-101

[68] Chou J, Franklin NR, Baeck SH, Jaramillo TF, McFarland EW. Gas-phase catalysis by micelle derived Au nanoparticles on oxide supports. Catalysis Letters. 2004;**95**:107-111

[69] Guczi L, Peto G, Beck A, Frey K, Geszti O, Molnar G, et al. Gold nanoparticles deposited on SiO_2/Si(100): Correlation between size, electron structure, and activity in CO oxidation. Journal of the American Chemical Society. 2003;**125**:4332-4337

[70] Somodi F, Borbáth I, Hegedűs M, Tompos A, Sajó IE, Szegedi A, et al. Modified preparation method for highly active Au/SiO_2 catalysts used in CO oxidation. Applied Catalysis A: General. 2008;**347**:216-222

[71] Gajan D, Guillois K, Delichere P, Basset JM, Candy JP, Caps V, et al. Gold nanoparticles supported on passivated silica: Access to an efficient aerobic epoxidation catalyst and the intrinsic oxidation activity of gold. Journal of the American Chemical Society. 2009;**131**:14667-14669

[72] Qureshi ZS, Sarawade PB, Hussain I, Zhu H, Al Johani H, Anjum DH, et al. Gold nanoparticles supported on fibrous silica nanospheres (KCC-1) as efficient heterogeneous catalysts for CO oxidation. ChemCatChem. 2016;**8**:1671-1678

[73] Yao X, Zhou S, Zhou H, Fan T. Pollen-structured hierarchically meso/macroporous silica spheres with supported gold nanoparticles for high-performance catalytic CO oxidation. Materials Research Bulletin. 2017;**92**:129-137

[74] Song Y, Chu X, Lin Y, Yang X. Pyrrolidone modifying gold nanocatalysts for enhanced catalytic activities in aerobic oxidation of alcohols and carbon monoxide. Journal of Chemistry. 2017;**5257296**:1-8. DOI: 10.1155/2017/5257296. Article ID 5257296

[75] Mikami Y, Dhakshinamoorthy A, Alvaro M, Garcia H. Catalytic activity of unsupported gold nanoparticles. Catalysis Science & Technology. 2013;**3**:58-69

[76] Huang Z, Li F, Chen B, Yuan G. Sustainable catalytic oxidation of alcohols over the interface between air and water. Green Chemistry. 2015;**17**:2325-2329

[77] Kucerová G, Strunk J, Muhler M, Behm RJ. Effect of titania surface modification of mesoporous silica SBA-15 supported Au catalysts: Activity and stability in the CO oxidation reaction. Journal of Catalysis. 2017;**356**:214-228

[78] Zhao D, Xu BQ. Enhancement of Pt utilization in electrocatalysts by using gold nanoparticles. Angewandte Chemie International Edition. 2006;**45**:4955-4959

[79] Chen JG, Menning CA, Zellner MB. Monolayer bimetallic surfaces: Experimental and theoretical studies of trends in electronic and chemical properties. Surface Science Reports. 2008;**63**:201-225

[80] Tai Y, Tajiri K. Preparation, thermal stability, and CO oxidation activity of highly loaded Au/titania-coated silica aerogel catalysts. Applied Catalysis A: General. 2008;**342**:113-118

[81] Tai Y, Yamaguchi W, Tajiri K, Kageyama H. Structures and CO oxidation activities of size-selected Au nanoparticles in mesoporous titania-coated silica aerogels. Applied Catalysis A: General. 2009;**364**:143-149

[82] Tai Y, Yamaguchi W, Okada M, Ohashi F, Shimizu K, Satsuma A, et al. Depletion of CO oxidation activity of supported Au catalysts prepared from thiol-capped Au nanoparticles by sulfates formed at Au–titania boundaries: Effects of heat treatment conditions on catalytic activity. Journal of Catalysis. 2010;**270**:234-241

[83] Peza-Ledesma CL, Escamilla-Perea L, Nava R, Pawelec B, Fierro JLG. Supported gold catalysts in SBA-15 modified with TiO_2 for oxidation of carbon monoxide. Applied Catalysis A: General. 2010;**375**:37-48

[84] Yan W, Petkov V, Mahurin SM, Overbury SH, Dai S. Powder XRD analysis and catalysis characterization of ultra-small gold nanoparticles deposited on titania-modified SBA-15. Catalysis Communications. 2005;**6**:404-408

[85] Ruszel M, Grzybowska B, Łaniecki M, Wojtowski M. Au/Ti-SBA-15 catalysts in CO and preferential (PROX) CO oxidation. Catalysis Communications. 2007;**8**:1284-1286

[86] Van den Berg MWE, Toni AD, Bandyopadhyay M, Gies H, Grünert W. CO oxidation with Au/TiO_2 aggregates encapsulated in the mesopores of MCM-48: Model studies on activation, deactivation and metal–support interaction. Applied Catalysis A: General. 2011;**391**:268-280

[87] Toni AD, Gies H, Grunert W. The impact of water on CO oxidation with Au/TiO_2 catalysts: Poison or promotor? A study with an Au–TiO_2/MCM-48 model catalyst. Catalysis Letters. 2011;**141**:1282-1287

[88] Horvath A, Beck A, Sarkany A, Stefler G, Varga Z, Geszti O, et al. Silica-supported Au nanoparticles decorated by TiO_2: Formation, morphology, and CO oxidation activity. Journal of Physical Chemistry B. 2006;**110**:15417-15425

[89] Gabaldon JP, Bore M, Datye AK. Mesoporous silica supports for improved thermal stability in supported Au catalysts. Topics in Catalysis. 2007;**44**(1-2):253-262

[90] Beck A, Horvath A, Stefler G, Scurrell MS, Guczi L. Role of preparation techniques in the activity of Au/TiO_2 nanostructures stabilised on SiO_2: CO and preferential CO oxidation. Topics in Catalysis. 2009;**52**:912-919

[91] Destro P, Kokumai TM, Scarpellini A, Pasquale L, Manna L, Colombo M, et al. The crucial role of the support in the transformations of bimetallic nanoparticles and catalytic performance. ACS Catalysis. 2018;**8**:1031-1037

[92] Destro P, Marras S, Manna L, Colombo M, Zanchet D. AuCu alloy nanoparticles supported on SiO_2: Impact of redox pretreatments in the catalyst performance in CO oxidation. Catalysis Today. 2017;**282**:105-110

[93] Bauer JC, Mullins D, Li M, Wu Z, Payzant EA, Overbury SH, et al. Synthesis of silica supported AuCu nanoparticle catalysts and the effects of pretreatment conditions for the CO oxidation reaction. Physical Chemistry Chemical Physics. 2011;**13**:2571-2581

[94] Liu X, Wang A, Li L, Zhang T, Mou CY, Lee JF. Structural changes of Au–Cu bimetallic catalysts in CO oxidation: In situ XRD, EPR, XANES, and FT-IR characterizations. Journal of Catalysis. 2011;**278**:288-296

[95] Escamilla-Perea L, Peza-Ledesma CL, Nava R, Rivera-Muñoz EM, Pawelec B, Fierro JLG. CO oxidation at 20°C over Au/SBA-15 catalysts decorated by Fe_2O_3 nanoparticles. Catalysis Communications. 2011;**15**:108-112

[96] Ramírez-Garza RE, Pawelec B, Zepeda TA, Martínez-Hernández A. Total CO oxidation over Fe-containing

Au/HMS catalysts: Effects of gold loading and catalyst pretreatment. Catalysis Today. 2011;**172**:95-102

[97] Zhang W, Lu X, Zhou W, Wu F, Li J. Mesoporous iron oxide-silica supported gold catalysts for low-temperature CO oxidation. Chinese Science Bulletin. 2014;**59**(31):4008-4013

[98] Liu X, Wang A, Yang X, Zhang T, Mou CY, Su DS, et al. Synthesis of thermally stable and highly active bimetallic Au-Ag nanoparticles on inert supports. Chemistry of Materials. 2009;**21**:410-418

[99] Wang AQ, Chang CM, Mou CY. Evolution of catalytic activity of Au-Ag bimetallic nanoparticles on mesoporous support for CO oxidation. Journal of Physical Chemistry B. 2005;**109**:18860-18867

[100] Yen CW, Lin ML, Wang A, Chen SA, Chen JM, Mou CY. CO oxidation catalyzed by Au-Ag bimetallic nanoparticles supported in mesoporous Silica. Journal of Physical Chemistry B. 2009;**113**:17831-17839

[101] Escamilla-Perea L, Nava R, Pawelec B, Rosmaninho MG, Peza-Ledesma CL, Fierro JLG. SBA-15-supported gold nanoparticles decorated by CeO_2: Structural characteristics and CO oxidation activity. Applied Catalysis A: General. 2010;**381**:42-53

[102] Liotta LF, Pantaleo G, Puleo F, Venezia AM. Au/CeO_2-SBA-15 catalysts for CO oxidation: Effect of ceria loading on physic-chemical properties and catalytic performances. Catalysis Today. 2012;**187**:10-19

[103] Chiang CW, Wang A, Mou CY. CO oxidation catalyzed by gold nanoparticles confined in mesoporous aluminosilicate Al-SBA-15: Pretreatment methods. Catalysis Today. 2006;**117**:220-227

[104] Chiang CW, Wang AQ, Wan BZ, Mou CY. High catalytic activity for CO oxidation of gold nanoparticles confined in acidic support Al-SBA-15 at low temperatures. Journal of Physical Chemistry B. 2005;**109**:18042-18047

[105] Xu X, Li J, Hao Z, Zhao W, Hu C. Characterization and catalytic performance of Co/SBA-15 supported gold catalysts for CO oxidation. Materials Research Bulletin. 2006;**41**:406-413

[106] Xu J, White T, Li P, He C, Yu J, Yuan W, et al. Biphasic Pd-Au alloy catalyst for low-temperature CO oxidation. Journal of the American Chemical Society. 2010;**132**:10398-10406

[107] Doherty RP, Krafft JM, Méthivier C, Casale S, Remita H, Louis C, et al. On the prmoting effect of Au on CO oxidation kinetics of Au–Pt bimetallic nanoparticles supported on SiO_2: An electronic effect? Journal of Catalysis. 2012;**287**:102-113

[108] Somodi F, Borbáth I, Hegedűs M, Sajó IE, Szegedi A, Rojas S, et al. Promoting effect of tin oxide on the activity of silica-supported gold catalysts used in CO oxidation. Reaction Kinetics and Catalysis Letters. 2008;**95**(1):175-183

[109] Bore MT, Mokhonoana MP, Ward TL, Coville NJ, Datye AK. Synthesis and reactivity of gold nanoparticles supported on transition metal doped mesoporous silica. Microporous and Mesoporous Materials. 2006;**95**:118-125

[110] Beck A, Horvath A, Stefler G, Katona R, Geszti O, Tolnai G, et al. Formation and structure of Au/TiO_2 and Au/CeO_2 nanostructures in mesoporous SBA-15. Catalysis Today. 2008;**139**:180-187

[111] Luo J, Ersen O, Chu W, Dintzer T, Petit P, Petit C. Anchoring and promotion effects of metal oxides on silica supported catalytic gold nanoparticles. Journal of Colloid and Interface Science. 2016;**482**:135-141

[112] Zepeda TA, Martinez-Hernandez A, Guil-Lopez R, Pawelec B. Preferential CO oxidation in excess of hydrogen over Au/HMS catalysts modified by Ce, Fe and Ti oxides. Applied Catalysis B: Environmental. 2010;**100**:450-462

Chapter 4

Evolution of Gold Nanoparticles in Radiation Environments

Samuel A. Briggs and Khalid Hattar

Abstract

Gold nanoparticles are being explored for several applications in radiation environments, including uses in cancer radiotherapy treatments and advanced satellite or detector applications. In these applications, nanoparticle interactions with energetic neutrons, photons, and charged particles can cause structural damage ranging from single atom displacement events to bulk morphological changes. Due to the diminutive length scales and prodigious surface-to-volume ratios of gold nanoparticles, radiation damage effects are typically dominated by sputtering and surface interactions and can vary drastically from bulk behavior and classical models. Here, we report on contemporary experimental and computational modeling efforts that have contributed to the current understanding of how ionizing radiation environments affect the structure and properties of gold nanoparticles. The future potential for elucidating the active mechanisms in gold nanoparticles exposed to ionizing radiation and the subsequent ability to predictively model the radiation stability and ion beam modification parameters will be discussed.

Keywords: radiation damage, gold nanoparticles, sputtering, radiation therapy, ion beam modification

1. Introduction

Ionizing radiation is known to have the ability to drastically alter material microstructure and performance primarily through the displacement of constituent atoms from their lattice sites, resulting in the generation of damage and point defects [1, 2]. These point defects tend to diffuse and coalesce into larger, ordered defect structures such as dislocation loops, cavities, and stacking faults, typically resulting in deleterious effects in structural materials such as radiation-induced hardening and embrittlement or void swelling [1, 2]. Irradiation can also cause solute redistribution in alloys and composite materials, encouraging the precipitation of secondary phases or promoting localized corrosion, which has been utilized in metal, ceramics, and polymers to provide added functionality not previously possible [3–8]. The fundamental mechanisms of radiation damage and the subsequent effects on materials properties and performance have been and continue to be heavily studied in commonly used and candidate materials for nuclear reactor and extraterrestrial applications.

While the potential effects of ionizing radiation environments are not commonly considered when discussing nanoparticle performance, there are a subset of both current and potential future applications for gold nanoparticles that

 IntechOpen

necessitate investigation and understanding of how their structure and properties change when exposed to energetic photons and particles. Radiation therapy is likely one of the most notable examples, in which gold nanoparticle injections in the vicinity of a tumor or conjugation to antibodies such that they preferentially bind to cancer cells have been shown to increase local radiation dose during X-ray irradiation through secondary photon interactions, thus lessening the absorbed dose in surrounding healthy tissue [9–13]. A similar local dose enhancement effect has been shown for proton and heavy ion irradiation therapies, which are being pursued to further reduce dose to healthy tissue as a consequence of how energetic charged particles deposit a majority of their energy at the end of their range [14–16]. Furthermore, potential applications in satellites or space electronics would involve bombardment with high-energy cosmic radiation [17, 18].

The mechanism of radiation damage in materials with limited dimensions has been shown to be significantly different from bulk material behavior [19–23]. Due to the prodigious surface-to-volume ratio of nanoparticles compared to even thin films, they have emerged as an attractive material choice for many applications. However, this same surface-to-volume ratio results in sputtering and free surface effects ultimately dominating the radiation response [20, 21]. Put simply, instead of diffusing into organized defect structures, the generated point defects tend to annihilate at the particle surface, or material is ejected from the particle volume, as a result of the energetic collision [24, 25]. This can drastically alter the shape of individual particles and cause agglomeration of closely spaced particle groupings [26–28]. Such morphological changes have the potential to adversely affect their efficacy in applications where the high surface-to-volume ratio and local structure are essential for performance.

This chapter attempts to summarize the current body of work relating to interactions of ionizing radiation with gold nanoparticles. Broader reviews of the interaction of nanostructured materials can be found in the detailed reviews by A.V. Krasheninnikov and K. Nordlund titled "Ion and electron irradiation-induced effects in nanostructured materials" [21] and the more recent review by X. Zhang et al. titled "Radiation damage in nanostructured materials" [29]. As one might expect, the interactions of radiation with elemental gold is highly dependent on the species of the incident radiation. Both the length scales of interaction and energy transferred per interaction event can vary widely due to the energy, mass, and charge of incident radiation, as highlighted in **Figures 1** and **2**. As such, this chapter

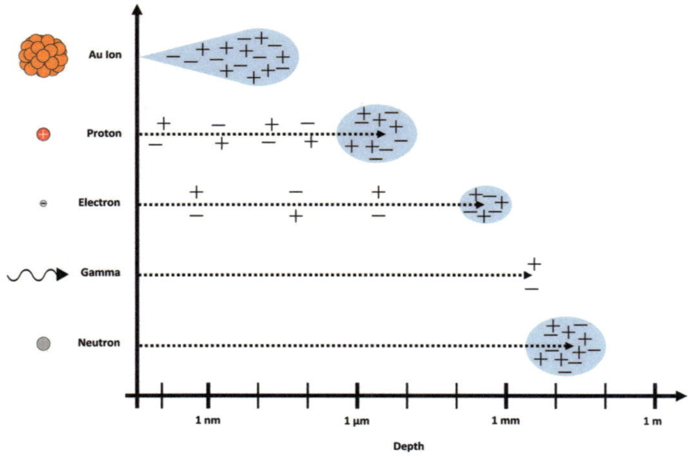

Figure 1.
Illustration of the relative average interaction length scales for various types of radiation at an energy of 1 MeV incident on gold. Dotted lines indicate the mean free path of the incident particles. Positive and negative signs indicate ionization events, while blue highlighted areas denote displacement damage cascades.

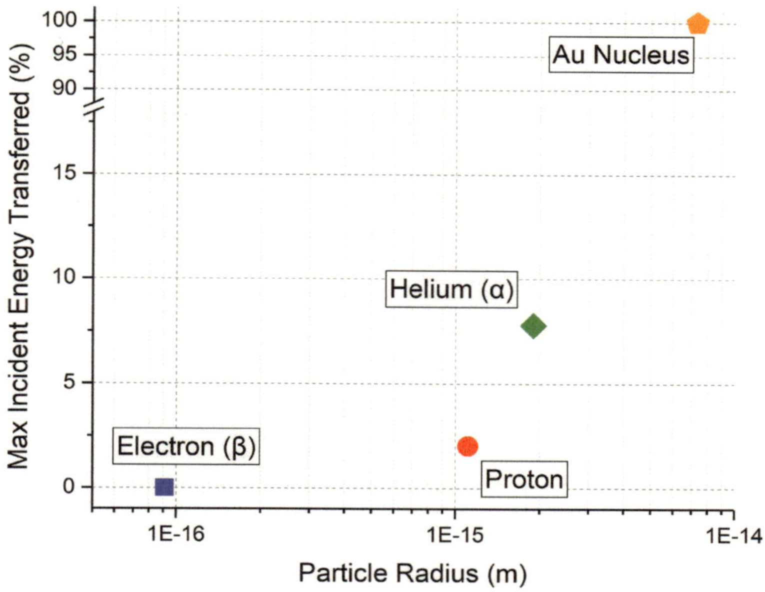

Figure 2.
Plot showing the theoretical maximum fraction of incident particle energy transferred via a ballistic collision of various particle types with a gold atom.

is organized according to the type of radiation environment a gold nanoparticle might encounter, and the subsequent sections highlight experimental studies of how the structure and properties of gold nanoparticles are altered by bombardment with specific types of energetic radiation. Section 2 discusses ionizing photon interactions, while Section 3 considers neutron environments. Sections 4–6 focus on different types of charged particle irradiation: beta particles (electrons/positrons), light ions (protons through alpha particles), and heavy ions. Section 6 also explores potential applications of ion beam modification in both freestanding and embedded gold nanoparticles from heavy ion bombardment. Finally, Section 7 discusses future experimental and modeling work needed to enhance our understanding of radiation effects in gold nanoparticles and speculates as to potential future applications combining gold nanoparticles with radiation environments.

2. Ionizing photon interaction with gold nanoparticles

The three primary interactions of photons with matter include the photoelectric effect, Compton scattering, and pair production [30]. In a photoelectric event, a photon completely transfers its energy to an orbital electron, ejecting it from its shell and ionizing the atom. This is the dominant interaction mechanism for low-energy ($E < \approx 0.5$ MeV) photons. In Compton scattering, an incident photon transfers a portion of its energy to an orbital electron, and both the electron and photon continue on, typically with different trajectories when compared to the incoming photon. This will be the dominant interaction mechanism for incident photons of intermediate energies (≈ 0.5 MeV $< E < \approx 2$ MeV). Finally, pair production dominates for high-energy photons and can only occur for incident photons greater than 1.022 MeV. In pair production, a photon interacts with the electromagnetic field surrounding the atomic nucleus and is converted into an electron and a positron, with the total photon energy less the rest energy of the two particles (0.511 MeV each) being shared between them. It should be noted that for high photon energies

(>≈8 MeV), photodisintegration can also occur, in which the incident photon causes the atom to emit one or more neutrons, potentially resulting in a radioactive isotope of gold [31].

Since incident photons interact primarily with the orbital electrons, the high atomic number (79) and density (19.32 g/cc) of gold significantly increase the likelihood of interaction compared to most other materials. The creation of secondary ionizing radiation through these photon interactions is the primary reason for the attractiveness of gold nanoparticles in radiation therapy applications [10]. Electrons and positrons resulting from the described interactions will lose their energy as Bremsstrahlung radiation (also known as braking radiation) as they slow down, and X-rays and Auger electrons will be produced as electrons shift from higher-energy orbitals to replace those ionized from lower-energy orbital positions. Furthermore, positrons generated from pair production will eventually annihilate with another electron, resulting in the generation of two 0.511 MeV photons. An illustration of these potential interaction mechanisms and resulting secondary radiation effects is shown in **Figure 3** [10]. This secondary radiation will typically deposit its energy in the vicinity of the gold particles, which can be taken advantage of to increase dose to cancerous cells [10, 11].

Photon irradiation experiments involving gold nanoparticles are prevalent, as X-ray therapies for cancer treatment are a fairly common practice [32–34]. While these studies are predominantly focused on the biological effects in the vicinity of cancerous tissue, they highlight how photons interact with these high atomic number (high-Z) nanoparticles to deposit their energy locally. High-energy photons are also expected to cause Frenkel pair displacement damage (i.e., vacancy and interstitial defect pairs), primarily through the energetic electrons that they tend to generate. However, the effects of these displacement effects on the long-term stability of nanoparticle structures, if any, have not been studied to our knowledge.

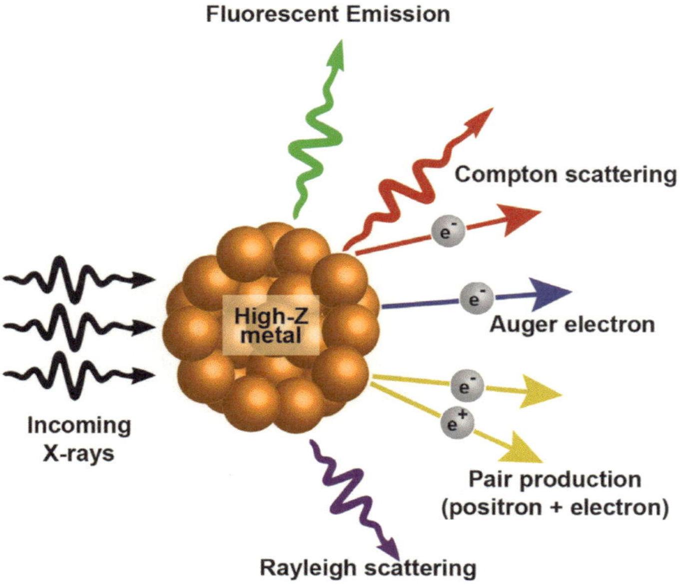

Figure 3.
Schematic illustrating potential interactions of incident photons with a gold atom or other high-Z materials [10]. Source: http://tcr.amegroups.com/article/view/1550/html. AME Publishing Company. Republished with permission of Pioneer Bioscience Publishing Company, from D. Kwatra et al., Translational Cancer Research, 2(4), pp. 332, 2013; permission conveyed through Copyright Clearance Center, Inc.

In addition to understanding the role of radiation on the nanoparticle and the surrounding material, a few research groups have shown that gamma and other forms of ionizing photons can even be used to induce nucleation and growth of nanoparticles out of solution including far-from-equilibrium structures that might be difficult to obtain through other more classical chemical synthesis routes [35, 36]. Most studies and reviews of the response of gold nanoparticles to various stimuli limit themselves to ionizing radiation produced by photons [37]. To expand these previous reviews and studies, the remainder of this chapter will focus on neutron, beta, alpha, and heavier charged particle irradiation effects on gold nanoparticles.

3. Neutron interaction with gold nanoparticles

To understand the interactions that can possibly occur between an energetic neutron and a gold nanoparticle, one must first understand the well-studied interaction of a neutron and a gold atom. As gold is monoisotopic, Au-197 is assumed as a target for the purposes of gauging probable interactions. Thus, neutron interactions with gold atoms are dominated either by radiative capture (n,γ) reactions or by scattering. Capture reactions are most prevalent for thermal neutron spectra ($E < 1$ eV), though there is a notable resonance absorption peak for energies near 4.9 eV and several other resonances between 60 eV and 2 keV [38]. This absorption reaction coincides with emission of a gamma ray with an energy between 4.78 and 6.52 MeV [39] and results in an Au-198 nucleus that subsequently decays to stable Hg-198 via beta particle emission with an energy release of 1.37 MeV and a half-life of 2.7 days. The reaction can also potentially result in two metastable states of Au-198 that emit 312 and 811 keV gamma rays with half-lives of 124 ns and 2.3 days, respectively.

Scattering interactions are more common for higher energy neutron spectra. Due to the large difference in mass between neutrons and gold nuclei, neutrons undergoing elastic scattering interactions (i.e., ballistic collisions) only transfer approximately 1% of their energy to the nucleus [1]. Inelastic scattering interactions, though less common, can transfer significantly more energy, leaving the nucleus in an excited state that results in gamma ray production. With sufficiently energetic incident neutrons, both scattering mechanisms are capable of generating gold primary knock-on atoms (PKAs) and generating displacement damage in the material microstructure.

No evidence of experimental work investigating neutron effects on individual particle structure or morphology was found, likely due to the challenges associated with working with neutron beams or research reactors and the high fluences required to induce appreciable damage. The mean free path (average distance traveled before interaction) of a neutron is much higher than that of a charged particle due to the lack of Coulombic interaction (see **Figure 2**), such that a vast majority of neutrons incident on a gold nanoparticle are not expected to result in a damage event. However, local injection of gold nanoparticles has recently been shown to enhance the effectiveness of neutron radiation therapies [40].

4. Beta particle interaction with gold nanoparticles

Charged particles are common by-products of radioactive decay and nuclear reactions and are a primary component of cosmic radiation. Accelerators and ion beams are also common ion sources used in both research and industry. For example, ions are frequently used in radiation damage experiments to simulate material microstructures resulting from neutron radiation exposure in nuclear reactor environments primarily because it is much less costly and can achieve similar damaged microstructures in a fraction of the time [41]. Accelerators also have a slew of other potential applications ranging from materials analysis (e.g., electron microscopes, Rutherford backscattering) to ion beam modification.

Beta (electron or positron) radiation either incident on or produced in gold will primarily lose its energy via ionization and Bremsstrahlung radiation. As mentioned in the introduction, Bremsstrahlung radiation occurs due to electron acceleration from interaction with an atomic nucleus and results in the production of a photon of energy equal to the energy lost by the electron. This interaction is again Z-dependent and is quite common for high-Z materials like gold [42]. While displacement or knock-on damage resulting from electrons is commonly observed in transmission electron microscopy experiments [43], the significant difference in the masses of an electron and a gold nucleus requires electron energies in excess of 1.35 MeV to create a single Frenkel pair [44]. In many cases the role of the electron beam in altering the nanoparticle stability during these studies is not a result of interaction with the gold atoms itself, but with the organic capping ligands, as a result the stability of the gold nanoparticles to beta radiation is often dictated by the organic capping agent chosen [45]. One of the many examples of gold nanoparticle sintering due solely to electron beam effects can be seen in **Figure 4** [68]. The details of the particle orientation, organic capping, support film, and electron beam condition will alter the sintering rate, particle reorientation, and possible grain boundary character formed during the sintering process. This area is well studied in the electron microscopy community and is not reviewed further in this chapter. Interested readers are referred to the newest edition of the classic Williams and Carter textbook, which contains significant additions discussing a range of electron beam effects [43].

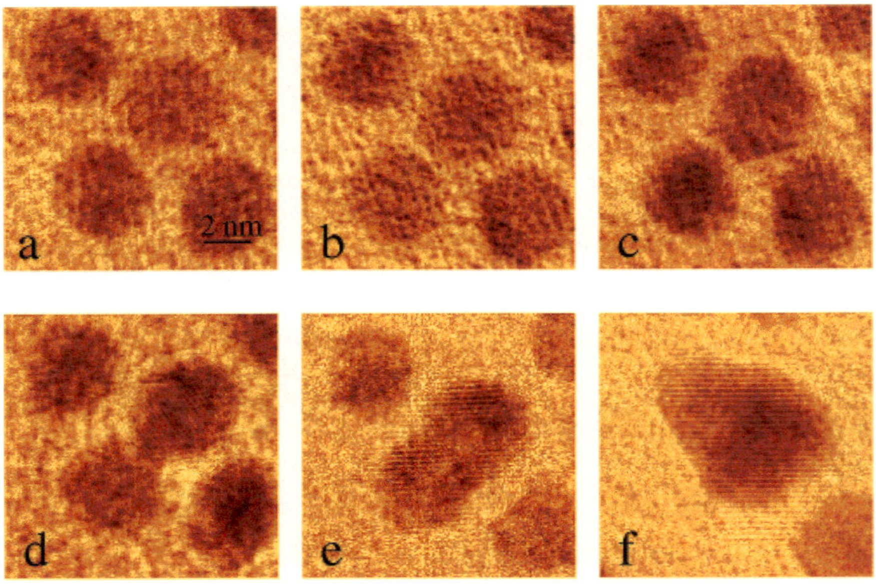

Figure 4.
Bright-field TEM images of dodecanethiol-passivated gold nanoparticles with core diameter 4.8 nm (a) before and after focused 200 keV electron beam irradiation with a dose of (b) 7.1, (c) 16.4, (d) 33.7, (e) 73.7, and (f) 149.8 μC/μm² [68]. Source: https://pubs.acs.org/doi/abs/10.1021/la0533157. American Chemical Society. Reprinted with permission from Y. Chen et al., Langmuir, Vol. 22, pp. 2851, 2006. Copyright 2006 American Chemical Society.

5. Light ion interaction with gold nanoparticles

For the purpose of this chapter, light ions will be defined as energetic particles as light as proton or as heavy as helium ions (alpha particles). Alpha particles are common products of radioactive decay for actinides and other heavy radioactive isotopes, but, along with protons, deuterons, and tritons, they can also result from and induce a number of different nuclear reactions. For example, irradiation of Au-197 with protons with energies of 4.5 MeV or higher can cause a (p,n) reaction resulting in the production of metastable Hg-197m [46]. However, threshold energies of these reactions for gold are, in most cases, sufficiently high and reaction cross sections sufficiently low that these types of interactions rarely occur in practical applications. More often, light ions will interact via ionization and through Coulombic forces. Similar to the other types of interactions discussed, ionization has the potential to result in the emission of characteristic X-rays and other secondary radiations. Coulombic interactions with other atomic nuclei can tend to cause displacement damage, usually in the form of Frenkel pairs or small, isolated cascades.

Proton irradiation experiments have commonly been conducted in the context of increasing local dose for proton therapy-based applications [16, 47, 48]. Again, the primary mechanism for this dose enhancement comes from local energy deposition from ionization and secondary radiation that is produced. He irradiation experiments are not typically performed, but the effect can be assumed to be similar. Evidence of the effects of proton and He irradiation on individual gold nanoparticle structure is lacking in the literature.

6. Heavy ion interaction with gold nanoparticles

Energetic heavy ions, which will be defined as ions heavier than a helium atom, can result from recoil following a nuclear reaction, decay, or fission events, though sources relevant to gold nanoparticle applications will often likely come from particle accelerators. Initiation of nuclear reactions with heavy ion irradiation is improbable, and energy is typically deposited via ionization and Coulombic interaction similar to their less massive counterparts. The primary distinction is that, due to their size, heavy ions are capable of transferring much more energy to the gold atoms they interact with, resulting in large displacement damage cascades (see **Figures 1** and **2**). Cascade clustering, in which several point defects are formed as a result of a single interaction event and coalesce into larger and less mobile defect structures, results in a much more disordered microstructure and production of a smaller fraction of freely migrating defects. Complex interactions between these point- and multi-defect structures can significantly affect their stability and mobility [49]. As alluded to in the introduction, the effects of limited dimensions in the formation of surface cascades and sputtering in gold have been well known for over two decades [50]. At these limited dimensions, the effect of viscous flow during ballistic interactions is thought to have a significant role [50, 51]. These and many other size effects resulting from heavy ion irradiation in gold nanoparticles will be explored in greater detail in the following subsections.

6.1 Radiation stability of freestanding gold nanoparticles

In general, the response of nanostructured materials to radiation damage is still poorly understood [29]. Despite the limited understanding in the general field, freestanding gold nanoparticles (usually drop casted onto carbon or silicon nitride TEM grids) have been used as the model system for testing and validation of TEM with in-situ ion irradiation capabilities [52–54]. Expanding on the known enhanced sputtering rates observed in gold thin foils exposed to a range of noble gas ions [55], it was later shown by Ilinov et al. that gold nanorods irradiated with 80 keV Xe demonstrated sputtering rates with three orders of magnitude higher than predicted by classical sputtering simulations [27]. These results have been verified multiple times in various facilities around the world. A detailed qualitative example can be seen in **Figure 5** [56]. By performing in-situ TEM experiments, individual nanoparticles, as well as the individual number of ion strikes on that particle, can be tracked for the duration of the experiment. This surprising set of results means that classical models and expectations no longer hold true when predicting the radiation response of gold nanoparticles.

These experimental results contradict the classic Monte Carlo-based simulations of sputtering effects [57]. A more catastrophic image of radiation damage in various sizes and morphologies of gold nanoparticles is predicted by molecular dynamic simulations of heavy ion irradiation [20, 26, 27, 56, 58]. An example of these types of simulations comparing the expected sputtering from a flat surface versus significantly increased sputtering from a nanoparticle can be seen in **Figure 6** [58]. The recent work has correlated the in-situ TEM observation of nanorod evolution and sputtering with molecular dynamic simulations that provide greater insight into the role of local crystal orientation on the effects of individual ion strikes relative to the nanoparticle orientation [26, 56]. The sputtering of the nanoparticles has also been tied to sintering of clustered gold nanoparticles, which is not surprising due to the work imparted into such a small volume. This is best seen when the particles are examined via electron tomography, as can be seen in **Figure 7** [28]. Despite the higher number of studies into the radiation stability of gold nanoparticles, few studies explore the effects of gold particle size and morphology or the myriad of irradiation environmental variables. One of the few systematic studies looked at the response of gold nanoparticles during self-ion

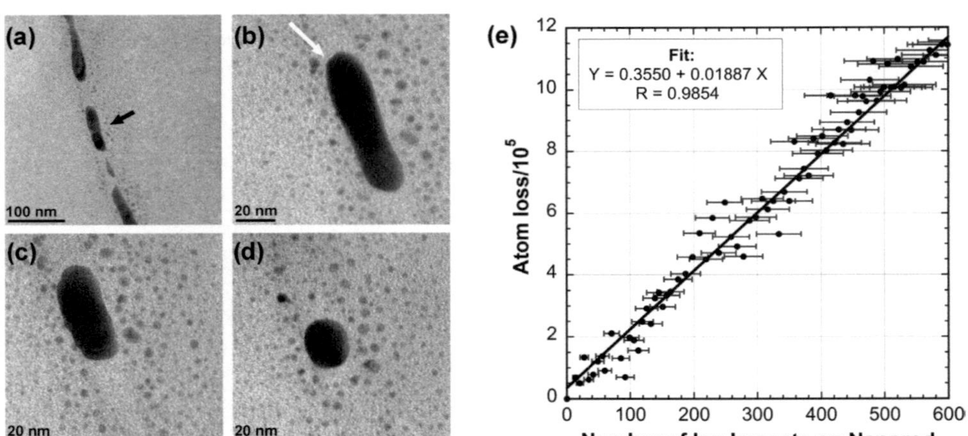

Figure 5.
Changes to Au nanowire due to irradiation with 80 keV Xe ions. (a) Segmentation due to "necking" and breaking of nanowire at grain boundaries following irradiation to a fluence of 2.1 × 10^{14} cm^{-2}, (b) nanorod at starting point for volume measurements—white arrow indicates projected direction of the ion beam which was incident at 60° to the specimen plane, (c) nanorod following irradiation to (additional) fluence of 1.6 × 10^{13} cm^{-2} (≈227 impacts on nanorod), (d) nanorod following irradiation to (additional) fluence of 5.5 × 10^{13} cm^{-2} (≈316 additional impacts on nanorod). All are bright-field TEM images and (e) plot of atom loss versus ion impacts for Au nanorod shown in panels (b)–(d) [56]. Source: https://journals.aps.org/prl/abstract/10.1103/PhysRevLett.111.065504. American Physical Society. Reprinted figure with permission from G. Greaves et al., Physical Review Letters, Vol. 111, pp. 065504-1, 2013. Copyright 2013 by the American Physical Society .

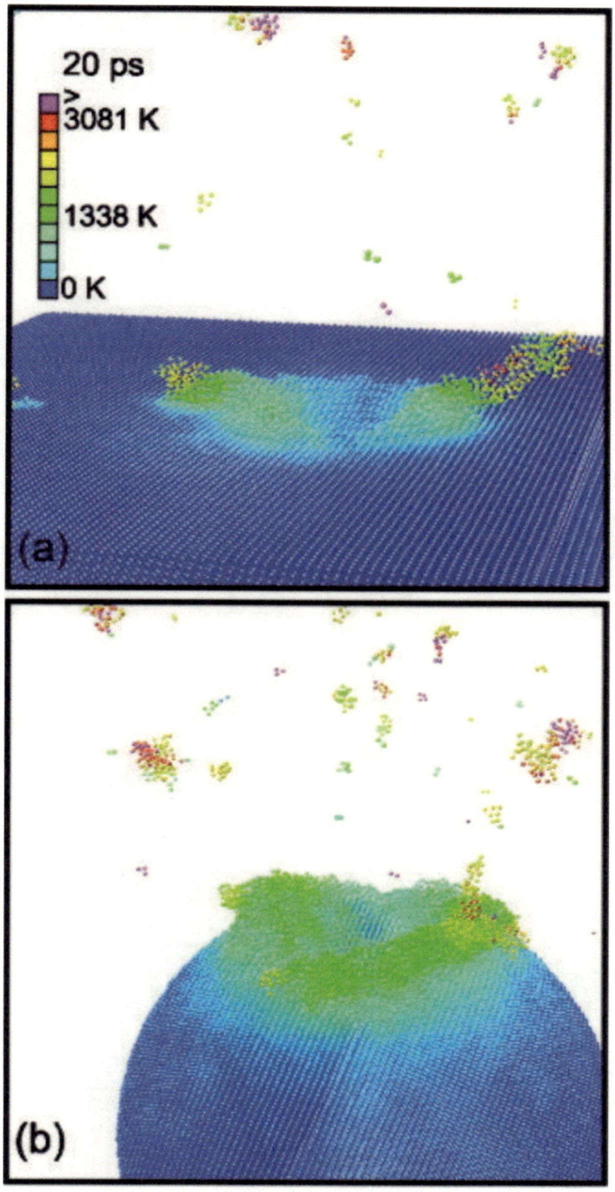

Figure 6.
Perspective view of 16 keV Au impact on (a) plane surface and on (b) nanoparticle at time = 20 ps after impact. Color denotes the local temperature [58]. Source: https://www.sciencedirect.com/science/article/pii/S1387380608000195. Elsevier B.V. Reprinted from International Journal of Mass Spectrometry, Vol. 272, S. Zimmermann and H.M. Urbassek, "Sputtering of nanoparticles: Molecular dynamics study of Au impact on 20nm sized Au nanoparticles," pp. 91-97, Copyright 2008, with permission from Elsevier.

irradiation. This study explored particles with average diameters of 5, 20, and 60 nm and altered the gold ion energy between 46 keV, 2.8 MeV, and 10 MeV [59]. It is very clear from the results presented in **Figure 8** that the response and stability of nominally spherical gold nanoparticles are heavily dependent on the order of magnitude changes in particle diameter, as the diameter approaches that of the cascade volume. Another heavy ion irradiation effect that has been noted at an even higher energy regime (956 MeV Pb) is that the nanoparticles can be ejected or desorb from the surface as a result of ion irradiation [60]. The combination of the

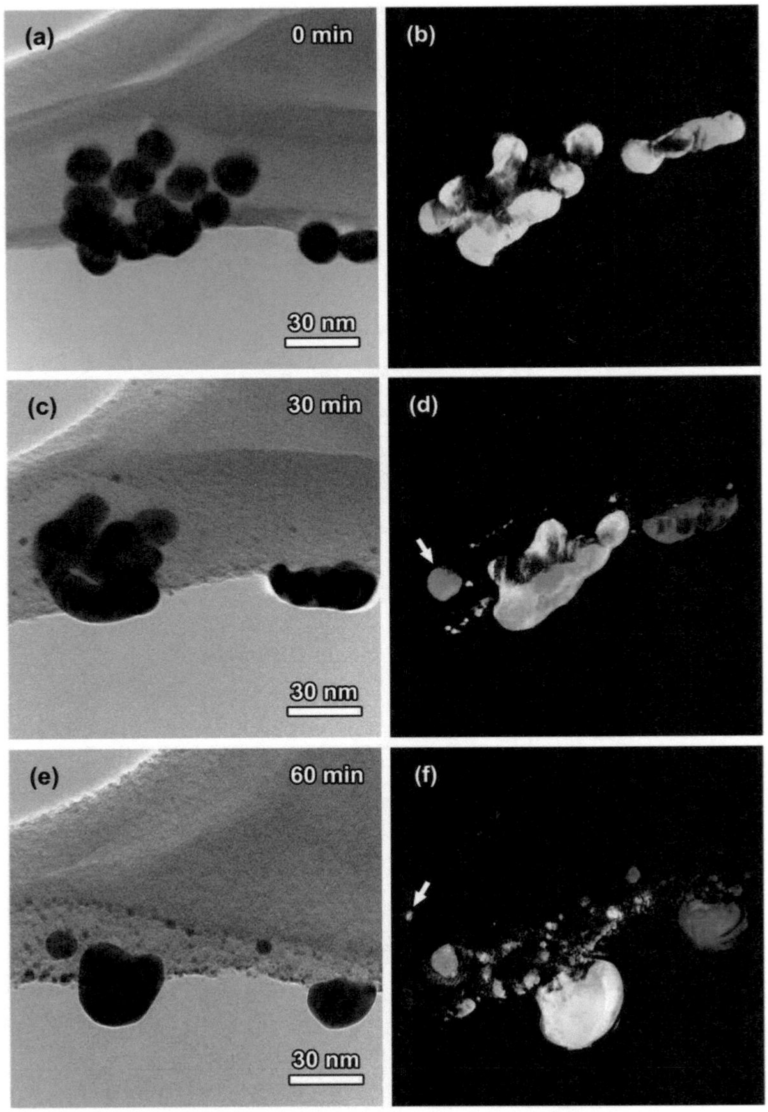

Figure 7.
Source electron micrographs and discrete 4D electron tomograms of Au nanoparticles irradiated with 3 MeV Cu^{3+}. (a, c, and e) Example micrographs from tilt series at increasing fluences up to ~10^{15} cm^{-2}. (b, d, f) Corresponding 3-D tomogram reconstructions, rotated to a different angle from the source micrograph. Source: http://pubs.rsc.org/en/content/articlelanding/2014/cc/c3cc49479a#!divAbstract. The Royal Society of Chemistry. Reproduced from [28] with permission of The Royal Society of Chemistry.

increased sputtering, sintering, ballistic destruction, and possible ejection demonstrates that the response of gold nanoparticles to heavy ion irradiation is far from that expected by classical theories and models based on bulk sample geometries. As such, significant further investigation is needed before any commercialization in displacement damage environments is considered.

6.2 Radiation stability of embedded gold nanoparticles

In addition to the work done on freestanding gold nanoparticles, there have been several studies exploring the response of heavy ion irradiation to particles

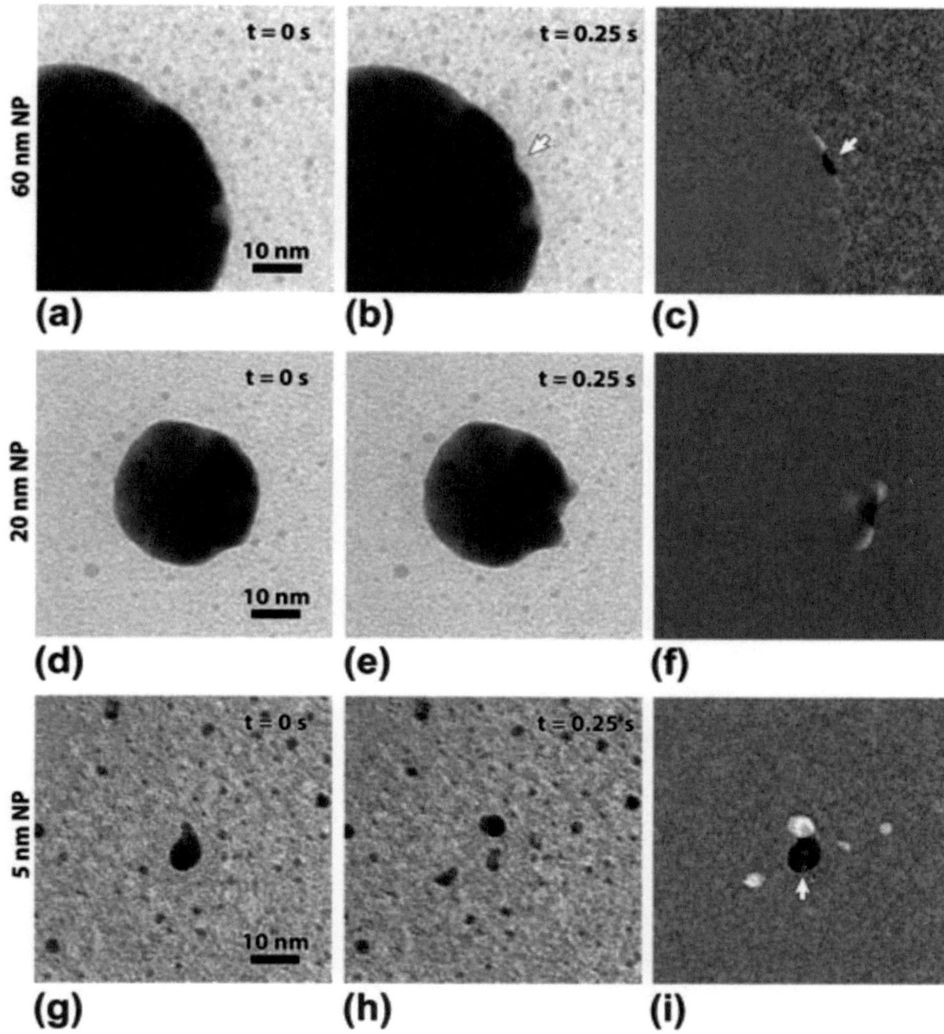

Figure 8.
Effects of single 46 keV ions in gold nanoparticles of decreasing size. Note that the magnification is similar for all micrographs. Each pair of micrographs is separated by one frame, about 0.25 s here. (a–c) A single ion strike in a 60-nm nanoparticle created a surface crater, marked by the white arrow. (c) The difference image highlights the change between (a) and (b); features present only in (a) are dark, and newly formed features present only in (b) appear light. (d–f) A single ion creating a crater in a 20-nm nanoparticle. (f) The difference image and (g–i) ~5-nm teardrop-shaped nanoparticle was initially surrounded by a number of previously sputtered particles. (h) The nanoparticle exploded, leaving several particles nearby. (i) Difference image showing the locations of the old and new particles. The white arrow indicates a fragment from (h) that is difficult to see in (i) because it overlapped the original nanoparticle location. Source: https://www.cambridge.org/core/journals/journal-of-materials-research/article/physical-response-of-gold-nanoparticles-to-single-self-ion-bombardment/F9933AF9ABAF6D1D3747AE1F8FFB5428. Materials Research Society. Ref. [59], reproduced with permission.

embedded in a matrix. The most common matrix that has been used for these studies is amorphous SiO_x, as it provides a stark contrast in composition, density, and properties to that of the gold nanoparticles. Similar to the observations in the freestanding nanoparticles, Rizza et al. [61–63] have shown that sputtering plays an important role in the evolution of the embedded particles. In addition to watching the size of the particles decrease, the matrix provides a medium that serves to slow down the travel of the sputtered gold resulting in a satellite structure of smaller gold nanoparticles surrounding the original. The example micrographs and size

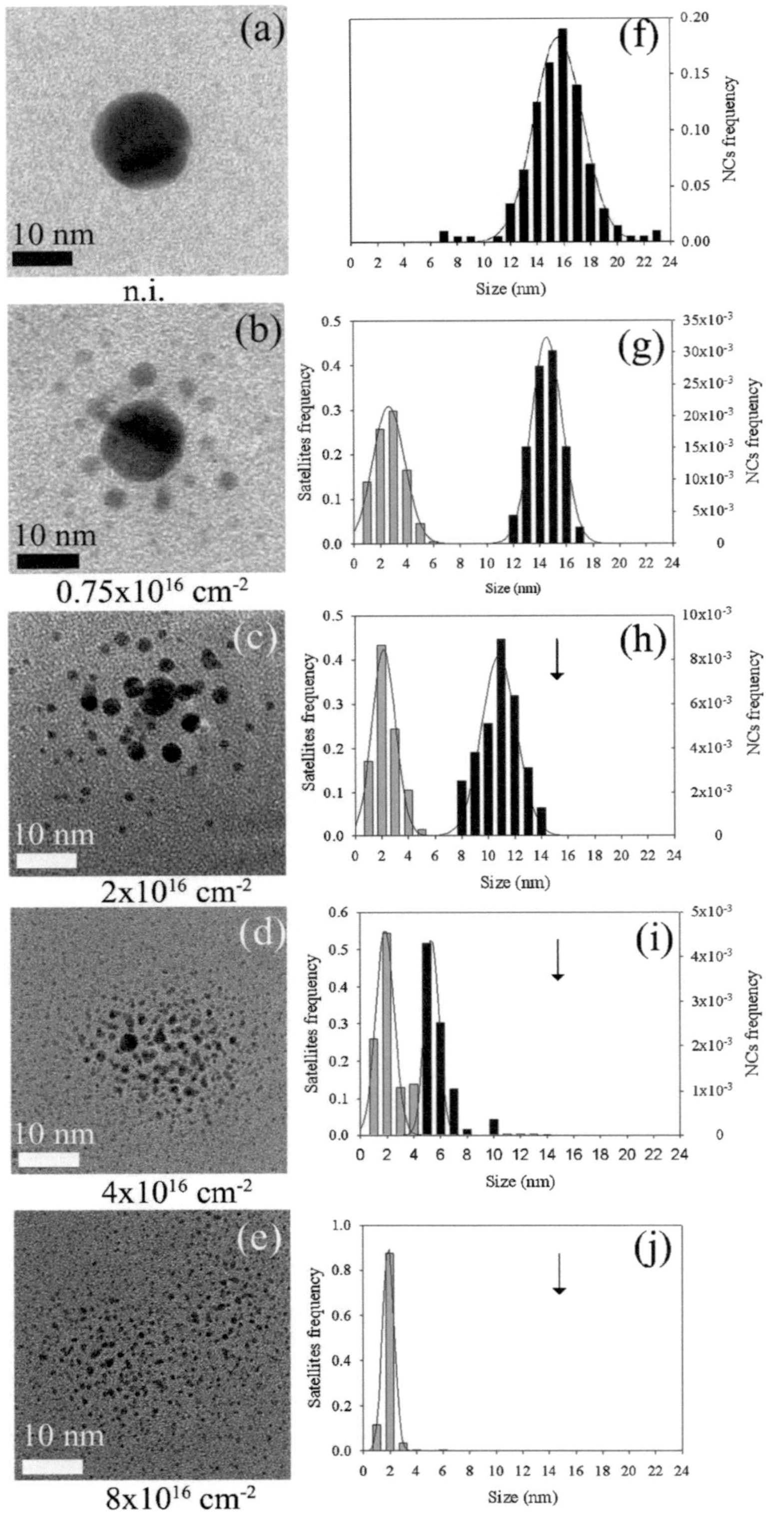

Figure 9.
(a–e) Bright-field TEM micrographs of the time sequence of an embedded nanoparticle evolution under 4 MeV Au irradiation at 300 K at increasing fluences up 8 × 10^{16} cm^{-2}. (f–j) The corresponding size distributions of nanoparticle and resulting satellites [61]. Source: https://aip.scitation.org/doi/abs/10.1063/1.2402351. American Institute of Physics. Reprinted from G. Rizza et al., Journal of Applied Physics, Vol. 101, pp. 014321, with the permission of AIP Publishing.

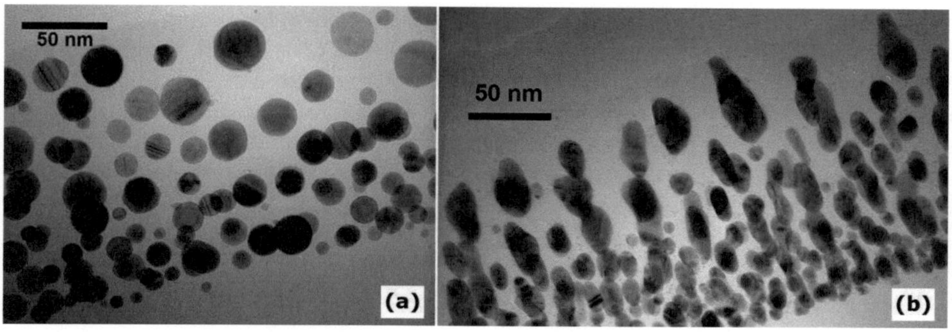

Figure 10.
(a) Cross-sectional transmission electron microscopy image of pristine film and (b) cross-sectional transmission electron microscopy image of an irradiated film at 45° beam incident normal with 120 MeV Au [64]. Source: https://aip.scitation.org/doi/10.1063/1.2764556. American Institute of Physics. Reprinted from Y.K. Mishra et al., Journal of Applied Physics, Vol. 91, pp. 063103, with the permission of AIP Publishing.

distributions of this transition from the original embedded nanoparticle through this satellite structure to the final clusters of smaller particles can be seen in **Figure 9** [61]. The shape of the particles can also be altered depending on the irradiation condition. Mishra et al. showed that by irradiating gold nanoparticles embedded in SiO_x with 120 MeV Au to a fluence of 3×10^{13} ions cm^{-2} with the ion beam tilted 45° off normal, elongated embedded nanoparticles can be formed, as seen in **Figure 10** [64]. There are probably many more very unique far-from-equilibrium structures that can originate from embedded gold nanoparticles by varying environmental parameters during the ion irradiation. These structures can be further controlled by combining the radiation damage of the embedded particle with thermal diffusion to study classical Ostwald ripening and other diffusional steps [65, 66]. The evolution of these particles can be understood through a combination of rate theory and Monte Carlo modeling [63].

In addition to the embedding of gold nanoparticles in a ceramic matrix, an effort has also been made to explore the radiation response of polymer matrices embedded with gold and other heavy metal nanoparticles. It has been shown that such materials can serve as easily processible and portable radiation shields [67]. With further study developing on the concepts presented in these examples of ion irradiated embedded gold nanoparticles, a range of complex far-from-equilibrium structures can be envisioned with an even greater number of potential novel applications.

7. Future directions

Clearly, the response of both freestanding and embedded gold nanoparticles is drastically different than gold in bulk or thin-film morphologies and is highly dependent on the radiation environment, particle morphology, and surface conditions. Additional work is needed to elucidate the underlying physics governing the increased sputtering and other scaling effects observed in gold nanoparticles [26, 27, 56]. Without a detailed understanding of the mechanisms active when the displacement damage length scale of the radiation event approaches that of the size of the nanoparticle exposed, it will be challenging to employ gold nanoparticles in most radiation environments. If the significant enhancement of sputtering is inherent and cannot be overcome, then the application of gold nanoparticles subject to ionizing radiation may be limited to those environments that produce minimal dose or sputtering yields. Conversely, if properly understood and controlled, the enhanced sputtering yields may also open new fields of study and possible applications utilizing the rapid degradation or the satellite morphology in the irradiated embedded nanoparticles seen in **Figure 8** [23, 61]. If not, the field may explore the response of other types of nanoparticles and nanostructured materials for inclusion in the next generation of devices that must withstand complex radiation environments. To be able to understand the effects of enhanced sputtering, systematic and thorough experimental and modeling efforts are needed along the lines of those presented in **Figure 6** [59].

8. Conclusion

Predicting the response of nanostructured materials for radiation environments is a new and rapidly developing field that is still poorly understood. The exposure to a range of ionizing photon irradiation has already found application in medical radiation therapies and will continue to gain traction in the coming years. In contrast, much less has been done to study neutron and charged particle irradiation effects on gold nanoparticles. The initial studies that have been done to study the radiation response of gold nanoparticles to charged particle irradiation indicate that a significant enhancement of sputtering yield is present. This enhanced sputtering leads to the rapid disintegration of the original nanoparticle and the formation of unique satellite nanoscale arrangements. Further modeling and experimental efforts are needed prior to the trusted incorporation of gold nanoparticles into other radiation environments (medical, space, or nuclear) being considered.

Acknowledgements

The authors would like to thank Prof. Jonathan Hinks for his insightful discussion and Ms. Macy Vereb for her assistance with the manuscript. This work was performed, in part, at the Center for Integrated Nanotechnologies, an Office of Science User Facility operated for the US Department of Energy (DOE) Office of Science. Sandia National Laboratories is a multi-mission laboratory managed and operated by the National Technology and Engineering Solutions of Sandia, LLC, a wholly owned subsidiary of Honeywell International, Inc., for the US DOE's National Nuclear Security Administration under contract DE-NA-0003525. The views expressed in the article do not necessarily represent the views of the US DOE or the US Government.

Conflict of interest

The authors declare that they have no competing interests or conflicts.

Author details

Samuel A. Briggs[1,2]* and Khalid Hattar[1]

1 Sandia National Laboratories, Albuquerque, NM, USA

2 Oregon State University, Corvallis, OR, USA

*Address all correspondence to: samuel.briggs@oregonstate.edu

References

[1] Was GS. Fundamentals of Radiation Materials Science. 2nd ed. New York: Springer-Verlag; 2017

[2] Rodriguez P, Krishnan R, Sundaram CV. Radiation effects in nuclear reactor materials—Correlation with structure. Bulletin of Materials Science. 1984;**6**(2):339-367

[3] Busby JT, Was GS, Kenik EA. Isolating the effect of radiation-induced segregation in irradiation-assisted stress corrosion cracking of austenitic stainless steels. Journal of Nuclear Materials. 2002;**302**:20-40

[4] Jiao Z, Was GS. Novel features of radiation-induced segregation and radiation-induced precipitation in austenitic stainless steels. Acta Materialia. 2011;**59**(3):1220-1238

[5] Was GS et al. Assessment of radiation-induced segregation mechanisms in austenitic and ferritic–martensitic alloys. Journal of Nuclear Materials. 2011;**411**(1-3): 41-50

[6] Mazzoldi P, Arnold GW. Ion Beam Modification of Insulators. Amsterdam (Netherlands): Elsevier; 1987

[7] Lee EH. Ion-beam modification of polymeric materials—Fundamental principles and applications. Nuclear Instruments and Methods in Physics Research Section B: Beam Interactions with Materials and Atoms. 1999;**151**(1-4):29-41

[8] Dearnaley G. Ion beam modification of metals. Nuclear Instruments and Methods in Physics Research Section B: Beam Interactions with Materials and Atoms. 1990;**50**(1-4):358-367

[9] Ferrari M. Cancer nanotechnology: Opportunities and challenges. Nature Reviews Cancer. 2005;**5**:161-171

[10] Kwatra D, Venugopal A, Anant S. Nanoparticles in radiation therapy: A summary of various approaches to enhance radiosensitization in cancer. Translational Cancer Research. 2013;**2**(4):330-342

[11] Liu Y et al. Metal-based nanoenhancers for future radiotherapy: Radiosensitizing and synergistic effects on tumor cells. Theranostics. 2018;**8**(7):1824-1849

[12] Rahman WN et al. Enhancement of radiation effects by gold nanoparticles for superficial radiation therapy. Nanomedicine: Nanotechnology, Biology and Medicine. 2009;**5**(2):136-142

[13] Stathakis S. The physics of radiation therapy. Medical Physics. 2010;**37**(3):1374-1375

[14] Kim JK et al. Therapeutic application of metallic nanoparticles combined with particle-induced X-ray emission effect. Nanotechnology. 2010;**21**(42):425102

[15] Lin Y et al. Biological modeling of gold nanoparticle enhanced radiotherapy for proton therapy. Physics in Medicine and Biology. 2015;**60**(10):4149-4168

[16] Polf JC et al. Enhanced relative biological effectiveness of proton radiotherapy in tumor cells with internalized gold nanoparticles. Applied Physics Letters. 2011;**98**(19):193702

[17] Holmes-Siedle A, Adams L. Handbook of Radiation Effects. New York, NY (United States): Oxford University Press Inc; 1993

[18] Ma T-P, Dressendorfer PV. Ionizing Radiation Effects in MOS Devices and Circuits. New York, NY (United States): Wiley-Interscience; 1989

[19] Järvi TT et al. Damage production in nanoparticles under light ion irradiation. Physical Review B. 2009;**80**(13):132101-1-132101-4

[20] Järvi TT, Nordlund K. Sputtering of freestanding metal nanocrystals. Nuclear Instruments and Methods in Physics Research Section B: Beam Interactions with Materials and Atoms. 2012;**272**:66-69

[21] Krasheninnikov AV, Nordlund K. Ion and electron irradiation-induced effects in nanostructured materials. Journal of Applied Physics. 2010;**107**(7):071301-1-071301-70

[22] Li J et al. In situ heavy ion irradiation studies of nanopore shrinkage and enhanced radiation tolerance of nanoporous Au. Scientific Reports. 2017;7:39484

[23] Knystautas E. Engineering Thin Films and Nanostructures with Ion Beams. Boca Raton, FL (United States): CRC Press; 2005

[24] Hoppe SM et al. Application of in-situ ion irradiation TEM and 4D tomography to advanced scintillator materials. In: SPIE 8509, Penetrating Radiation Systems and Applications XIII, 85090F; 19 October 2012

[25] Donnelly S, Birtcher R. Heavy ion cratering of gold. Physical Review B. 1997;**56**(21):13599

[26] Hinks JA et al. Effects of crystallographic and geometric orientation on ion beam sputtering of gold nanorods. Scientific Reports. 2018;**8**(1):512

[27] Ilinov A et al. Sputtering yields exceeding 1000 by 80 keV Xe irradiation of Au nanorods. Nuclear Instruments and Methods in Physics Research Section B: Beam Interactions with Materials and Atoms. 2014;**341**:17-21

[28] Bufford DC et al. In situ TEM ion irradiation and implantation effects on Au nanoparticle morphologies. Chemical Communications. 2014;**50**(57):7593-7596

[29] Zhang X et al. Radiation damage in nanostructured materials. Progress in Materials Science. 2018;**96**:217-321

[30] Martin JE. Physics for Radiation Protection. Weinheim, Germany: WILEY-VCH; 2006

[31] Ermakov AN et al. Multineutron photodisintegration of the Au-197 nucleus behind the giant dipole resonance. Physics of Atomic Nuclei. 2008;**71**:397

[32] Hainfeld JF, Slatkin DN, Smilowitz HM. The use of gold nanoparticles to enhance radiotherapy in mice. Physics in Medicine and Biology. 2004;**49**(18):N309-N315

[33] Mesbahi A. A review on gold nanoparticles radiosensitization effect in radiation therapy of cancer. Reports of Practical Oncology and Radiotherapy. 2010;**15**(6):176-180

[34] Shrestha S et al. Gold nanoparticles for radiation enhancement in vivo. Jacobs Journal of Radiation Oncology. 2016;**3**(1):026

[35] Gachard E et al. Radiation-induced and chemical formation of gold clusters. New Journal of Chemistry. 1998;**22**(11):1257-1265

[36] Nenoff TM et al. Radiological control of gold octahedral and prism nanoparticles. MRS Online Proceedings Library Archive. 2007;**1056**:1056-HH02-07

[37] Daniel M-C, Astruc D. Gold nanoparticles: Assembly, supramolecular chemistry, quantum-size-related properties, and applications toward biology, catalysis, and nanotechnology. Chemical Reviews. 2004;**104**(1):293-346

[38] Chadwick MB et al. ENDF/B-VII.1: Nuclear data for science and technology: Cross sections, covariances, fission product yields and decay data. Nuclear Data Sheets. 2011;**112**:2887

[39] Wasson OA et al. Au197(n,γ)Au198 reaction mechanism. Physical Review. 1968;**173**(4):1170-1184

[40] Kim EH et al. Gold nanoparticles as a potent radiosensitizer in neutron therapy. Oncotarget. 2017;**8**(68):112390-112400

[41] Was GS et al. Emulation of reactor irradiation damage using ion beams. Scripta Materialia. 2014;**88**:33-36

[42] Hippler R et al. Z dependence of bremsstrahlung radiation from free atoms. Physical Review Letters. 1981;**46**(25):1622-1625

[43] Williams DB, Carter CB. Transmission Electron Microscopy: A Textbook for Materials Science. New York: Springer; 2009

[44] Bauer W, Sosin A. Point defect studies in gold by electron irradiation at low temperatures. I. Threshold displacement energy and displacement cross section. Physical Review. 1964;**135**(2A):A521-A526

[45] Egerton R, Li P, Malac M. Radiation damage in the TEM and SEM. Micron. 2004;**35**(6):399-409

[46] Elmaghraby EK et al. Production of the mercury-197 through proton induced reaction on gold. Applied Radiation and Isotopes. 2010;**68**(9):1694-1698

[47] Kwon J et al. Dose distribution of electrons from gold nanoparticles by proton beam irradiation. International Journal of Medical Physics, Clinical Engineering and Radiation Oncology. 2015;**04**(01):49-53

[48] Li S et al. LET-dependent radiosensitization effects of gold nanoparticles for proton irradiation. Nanotechnology. 2016;**27**(45):455101

[49] Han W et al. Design of radiation tolerant materials via interface engineering. Advanced Materials. 2013;**25**(48):6975-6979

[50] Birtcher RC, Donnelly SE. Plastic flow induced by single ion impacts on gold. Physical Review Letters. 1996;**77**(21):4374-4377

[51] Ghaly M, Averback RS. Effect of viscous flow on ion damage near solid surfaces. Physical Review Letters. 1994;**72**(3):364-367

[52] Hinks J. A review of transmission electron microscopes with in situ ion irradiation. Nuclear Instruments and Methods in Physics Research Section B: Beam Interactions with Materials and Atoms. 2009;**267**(23-24):3652-3662

[53] Hinks JA. Transmission electron microscopy with in situ ion irradiation. Journal of Materials Research. 2015;**30**(9):1214-1221

[54] Hattar K, Bufford DC, Buller DL. Concurrent in situ ion irradiation transmission electron microscope. Nuclear Instruments and Methods in Physics Research Section B: Beam Interactions with Materials and Atoms. 2014;**338**:56-65

[55] Birtcher RC, Donnelly SE, Schlutig S. Nanoparticle ejection from gold during ion irradiation. Nuclear Instruments and Methods in Physics Research Section B: Beam Interactions with Materials and Atoms. 2004;**215**(1-2):69-75

[56] Greaves G et al. Enhanced sputtering yields from single-ion impacts on gold nanorods. Physical Review Letters. 2013;**111**(6):065504

[57] Ziegler JF, Ziegler MD, Biersack JP. SRIM—The stopping and range of ions in matter (2010). Nuclear Instruments and Methods in Physics Research Section B: Beam Interactions with Materials and Atoms. 2010;**268**(11-12):1818-1823

[58] Zimmermann S, Urbassek HM. Sputtering of nanoparticles: Molecular dynamics study of Au impact on 20 nm sized Au nanoparticles. International Journal of Mass Spectrometry. 2008;**272**(1):91-97

[59] Bufford DC, Hattar K. Physical response of gold nanoparticles to single self-ion bombardment. Journal of Materials Research. 2014;**29**(20):2387-2397

[60] Baranov I et al. Desorption of gold nanoclusters (2-150 nm) by 1 GeV Pb ions. Nuclear Instruments and Methods in Physics Research Section B: Beam Interactions with Materials and Atoms. 2005;**230**(1-4):495-501

[61] Rizza G et al. Ion beam irradiation of embedded nanoparticles: Toward an in situ control of size and spatial distribution. Journal of Applied Physics. 2007;**101**(1):014321-1-014321-7

[62] Rizza G et al. Chemically synthesized gold nanoparticles embedded in a SiO_2 matrix: A model system to give insights into nucleation and growth under irradiation. Physical Review B. 2007;**76**(24):245414

[63] Rizza G et al. Ion irradiation of gold inclusions in SiO_2: Experimental evidence for inverse Ostwald ripening. Nuclear Instruments and Methods in Physics Research Section B: Beam Interactions with Materials and Atoms. 2001;**178**(1-4):78-83

[64] Mishra YK et al. Synthesis of elongated Au nanoparticles in silica matrix by ion irradiation. Applied Physics Letters. 2007;**91**(6):063103-1-063103-3

[65] Ramjauny Y et al. Controlling the size distribution of embedded Au nanoparticles using ion irradiation. Journal of Applied Physics. 2010;**107**(10):104303-1-104303-8

[66] Pivin J et al. Diffusion process of metals in silica during ion irradiation. EPL (Europhysics Letters). 1997;**39**(6):623

[67] Nambiar S, Yeow JT. Polymer-composite materials for radiation protection. ACS Applied Materials & Interfaces. 2012;**4**(11):5717-5726

[68] Chen Y, Palmer RE, Wilcoxon JP. Sintering of passivated gold nanoparticles under the electron beam. Langmuir. 2006;**22**(6):2851-2855

Section 2

State-of-the-Art Catalysis

Chapter 5

Detoxification of Carcinogenic Dyes by Noble Metal (Ag, Au, Pt) Impregnated Titania Photocatalysts

Sivakumar Thiripuranthagan and Valentine Rupa

Abstract

Textile industries produce large quantities of waste water which is notoriously known to contain strong colour, suspended solids and COD. Although several treatment processes, such as filtration, coagulation, oxidation, ozonation, reverse osmosis, etc., have been practiced for many decades, they all have inherent limitations. They transfer the harmful pollutants from one phase to another and do not address detoxification. Semiconductor photocatalysis is a promising technique for photodegradation of hazardous chemicals found in waste waters. Among various semiconductors, TiO_2 has been studied widely. However, the main drawbacks associated with TiO_2 are (i) large band gap (Eg > 3.2 eV) (only UV active) and (ii) recombination of excitons. Dye sensitisation, coupling of semiconductors, and transitional metal doping are some of the methods reported to shift its optical response to visible region. Herein, nanoparticles of different noble metals such as Ag, Au and Pt were deposited on synthesised TiO_2, characterized by XRD, TEM, FT-IR, BET, UV-Vis and AAS and were subjected to the degradation of some textile dyes namely Tartrazine (TAZ), Reactive Yellow-17 (RY-17) and Reactive Black-5 (RB-5) under both UV and visible irradiations. The reaction conditions such as catalyst concentration, dye concentration, pH, irradiation time, light intensity, and additives were optimized for complete decolourisation and discussed.

Keywords: photocatalyst, gold impregnated titania, textile effluent, detoxification, carcinogenic dyes

1. Introduction

Dyeing and finishing processes are the two important steps in textile industries. Large number of synthetic dyes are used extensively in the textile dyeing process and these industries produce large quantities of waste water containing carcinogenic dyes. When ground water mixes with such waste water it gets polluted and cause allergy, damage of eye, brain, liver and reproductive organs and deformities in babies during pregnant period. Thus, water pollution is a major concern in the developing nations. Though dyes are aesthetic pollutants by nature of their colour, they may interfere with light penetration in the receiving water bodies thereby disturbing the biological processes.

Among the various types of dyes, reactive dyes are extensively used for dyeing cotton fabrics and hence they add a lot to water pollution. The volume of textile effluent discharged from dyeing industries mainly using reactive dyes are approximately calculated to be 3,00,000–4,00,000 L ton^{-1} fabric materials in a year [1].

The increased public concern and the stringent international environmental standards have prompted the need to develop novel treatment methods for converting dye effluents to harmless compounds. Although some conventional treatment technologies such as filtration, coagulation, flocculation etc. have been tried in the past they are not viable and green technologies. Zero generation of sludge and complete mineralisation of dyes are the two important criteria for the technology to be economically attractive and environmentally benign. In this context the present study of dye treatment by using photocatalysts which mineralises the carcinogens into CO_2, H_2O and mineral salts assumes great significance.

The ideal photocatalyst should be stable, inexpensive, non-toxic and highly photoactive. Another primary criteria for the degradation of organic compounds is that the redox potential of the $H_2O/OH^{\cdot}$ couple lies within the band gap of the semiconductor [2]. Several semiconductors have band gap energies sufficient for catalysing wide range of chemical reactions. These include TiO_2, WO_3, $SrTiO_3$, ZnO, ZnS, etc. Binary metal sulphide semiconductors such as CdS, CdSe and PbS are not sufficiently stable catalysts in aqueous media as they readily undergo photoanodic corrosion [3]. These materials are also known to be toxic. The iron oxides are not suitable semiconductors as they readily undergo photocathodic corrosion [4]. ZnO is unstable in water as $Zn(OH)_2$ being formed on the particle surface. This results in catalyst deactivation [5]. However, the anatase form of TiO_2 seems to be the best. The photoactivity of TiO_2 is known for approximately 60 years and investigated extensively. Although TiO_2 absorbs only approximately 5% of the solar light reaching the surface of the earth, it is the best-investigated semiconductor in the field of chemical conversion and storage of solar energy. Accordingly, many efforts have been made to sensitise titania for visible light induced photocatalytic reactions. Such sensitization techniques include (i) Doping with altervalent ions, (ii) heterojunctioning with other semiconductors, (iii) dye sensitization (iv), deposition of noble metals over semiconductors etc.

Herein we report the deposition of noble metals such as Au, Pt and Pd over synthesised titania. Noble metal is important because of their own catalytic activity and they actually modify the photocatalytic properties of the semiconductor by changing the distribution of electrons. Further, it can enhance the yield of a particular product or the rate of the photocatalytic reaction. The addition of a metal to a semiconductor surface also changes the reaction products. On the other hand, the loading level is important in governing the net effect of metallation as heavy metal loading induces faster electron-hole recombination [6]. One of the main reasons for the numerous studies of titania-supported metal catalysts is the so-called strong metal-support interaction (SMSI). The platinised titania exhibits a much higher activity in a series of photocatalytic reactions than pure anatase. Combined $Pt\text{-}RuO_2/TiO_2$ catalysts are the most promising ones for photocatalytic water cleavage [7]. The Au, Ag and Pd impregnated titania catalysts were synthesised, characterised by various instrumental techniques and evaluated towards the decolourisation of one azo dye tartrazine (TAZ) and two reactive dyes Reactive Yellow and Reactive Black (RY-17 and RB-5) (**Table 1**).

DOI: http://dx.doi.org/10.5772/intechopen.80467

Name of the dye	Molecular Formula	Structure
Tartrazine (TAZ) Molecular Weight-534.3 λ_{max}(nm)- 455 Dye Type- Azo C I No- C.I.19140	$C_{16}H_9N_4Na_3O_9S_2$	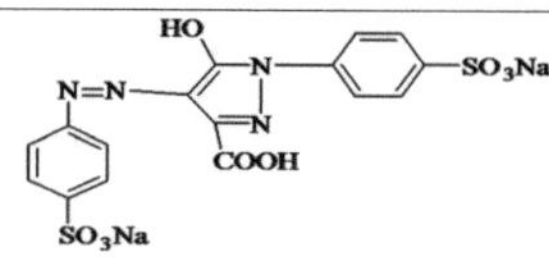
ReactiveYellow-17(RY-17) Molecular Weight-691.1 λ_{max}(nm)- 420 Dye Type- Reactive C.I No18852	$C_{21}H_{17}N_4K_2O_{10}S_3$	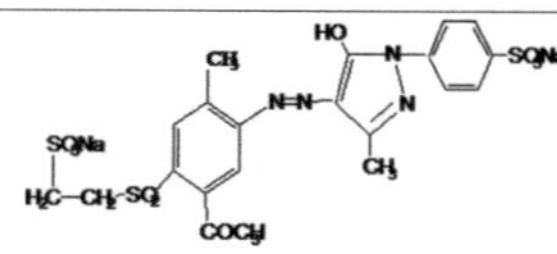
Reactive Black-5 (RB-5) Molecular Weight-927.4 λ_{max}(nm)- 597 Dye Type-Reactive C.I.20502	$C_{26}H_{21}N_5Na_4O_{19}S_6$	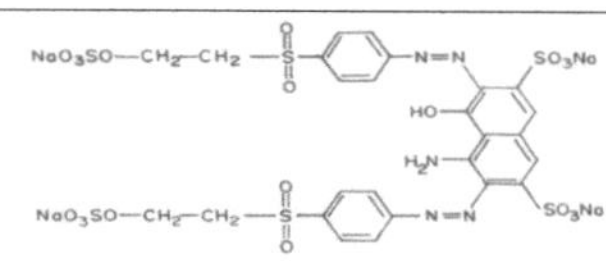

Table 1.
Physicochemical properties of various dyes under study.

2. Materials and methods

2.1 Materials

Textile dyes namely Tartrazine (TAZ), Reactive Yellow-17 (RY-17) and Reactive Black (RB-5) were procured from Bagmul Sons, India and was used without any further purification.

The anatase form of TiO_2 (P-25 Degussa) with particle size 30 nm and surface area 50 $m^2\ g^{-1}$ was used as such. The precursors for titania, Ag, Au and Pt were titanium isopropoxide (Lancaster >99% pure), Silver nitrate (Merck >99% pure), Chloroauric acid (Merck >99% pure) and tetraammineplatinum (II) chloride (Merck >99% pure) were used as received.

All other chemicals namely NaCl, C_2H_5OH, Na_2CO_3, H_2O_2, $K_2S_2O_8$, NH_4OH, H_2SO_4, isopropanol and acetic acid were obtained from Merck (purity >98%) and were used as received. The water employed for the studies was doubly distilled.

2.2 Synthesis of photocatalysts

The catalytic materials used in the present study were synthesised as described below:

2.2.1 Synthesis of TiO_2

Sol gel process was adopted for the synthesis of TiO_2. Titanium isopropoxide, isopropanol, water and acetic acid were used as starting materials. Solution A contained 17 mL titanium isopropoxide and 40 mL of isopropanol and Solution B contained 60 mL of isopropanol, 15 mL acetic acid and 5 mL water. Solutions A and B were mixed and stirred for 2 h. The formed TiO_2 sol was aged to get the gel of TiO_2. The obtained gel was dried, ground with mortar and pestle and finally calcined at 500°C for 3 h [8].

2.2.2 Synthesis of M/TiO_2 (M = Ag, Au and Pt)

The M/TiO_2 (M = Ag, Au and Pt) catalysts were prepared by photoreduction method [9]. Calculated amounts of noble metal precursors (0.14 g of $AgNO_3$ for

1% Ag/TiO_2, 0.15 g of $AuCl_3$ for 1% Au/TiO_2, 0.14 g of $[Pt(NH_3)_4]Cl_2$ for 1% Pt/TiO_2) taken in minimum amount of water was added to 10 g of synthesised TiO_2 and was stirred at 70°C till the evaporation of the solvent. The colloidal solution was then irradiated with 125 W halogen lamp until complete reduction was observed visually by the formation of black precipitate for Ag/TiO_2, Pt/TiO_2 and purple coloured product for Au/TiO_2.

2.3 Characterisation of the catalysts

2.3.1 X-ray diffraction analysis

Synthesised TiO_2 and M/TiO_2 photocatalyst were subjected to powder X-ray diffraction to confirm the crystalline size and the presence of the metal ion. The powder X-ray diffractograms were recorded using a X-ray diffractometer (PANalytical X'Pert Pro) with Cu Kα source having wavelength of 1.54 Å operating at 20 mA and 50 kV. The samples were scanned from 0.5–80° (2θ) for the confirmation of phase and presence of metal particles. The diameter of the metal particles was calculated from the (111) reflection of metal atom using Debye Scherrer equation,

$$d = K\lambda/\beta \cos\theta \qquad (1)$$

where d, diameter of the metal particle; K, constant; λ, wavelength of the X-ray (1.54 Å); β, full width at half maximum; θ, angle of diffraction.

2.3.2 Transmission electron microscopy

Particle size of the noble metals deposited on TiO_2 was determined from electron micrographs taken by using transmission electron microscope (JEOL JEM-2010F) operating at 200 kV. A drop of alcoholic solution of the catalysts was placed on Cu grid and kept aside for about 45 s. Then transmission electron micrographs were recorded for the particles present on the grid. From the TEM images, average particle size of the catalysts was determined be taking minimum of 30 particles.

2.3.3 BET surface area analysis

BET surface area analyser (Nova-Quantachrome 4200 e) was used to measure the surface area of all the catalysts. The measurement was done by using N_2 as probe molecule at liquid nitrogen temperature (−190°C). Before the measurement, all the catalysts were degassed at 300°C at the pressure of 10^{-5} Torr for 6 h. Linear portion of the BET plots was used for the determination of BET surface area.

2.3.4 Atomic absorption spectroscopic studies

The actual amount of metal content present in the synthesised catalysts (M/TiO_2, M = Ag, Au and Pt) was determined using atomic absorption spectrophotometer (Perkin Elmer 2380) after digesting the photocatalysts in aqua regia.

2.3.5 UV-visible spectrophotometry

UV-visible spectra of all the samples were recorded using UV-Vis double beam spectrophotometer (Hitachi U-2000) in the range 200–800 nm. The course of reduction of noble metal ions on TiO_2 particles and extent of decolourisation were studied using this technique. Further, the band gap energies (E_g) of the synthesised catalysts were calculated according to the equation

$$E_g = h\,c/\lambda \tag{2}$$

where h, Planck's constant; c, velocity of light (m/s); λ, wavelength (nm).

2.4 Photodecolourisation studies

2.4.1 Preparation of synthetic dye samples

Stock solutions (10^{-4} M for TAZ and 10^{-5} M for both RY-17 and RB-5) of the dye samples were prepared for the present study.

2.4.2 Batch photocatalytic reactor

The photocatalytic reactor was made up of quartz having dimensions 30 × 3 cm (height × diameter) provided with water circulation arrangement to maintain the temperature. The top portion of the reactor has ports for sampling.

The irradiation was carried out using low pressure mercury arc lamp (wave length 254 nm) as UV source and tungsten lamp (365 nm) as visible light source built into a lamp housing with polished anodised aluminium reflectors placed 6.5 cm away from the lamps. The entire reactor system was cooled using an inbuilt fan set up.

2.4.3 Photocatalytic reaction

250 mL of the desired dye sample was taken in a photocatalytic reactor. Calculated amount of TiO_2 (P-25 Degussa) or synthesised TiO_2 or M/TiO_2 was added to the photocatalytic reactor and the reaction mixture was magnetically stirred before and during illumination. After specific time interval of irradiation, suitable aliquots of the sample was withdrawn and analysed to find out the extent of decolourisation and degradation.

2.4.4 UV-vis analysis of the decolourised products

The extent of photocatalytic decolourisation of dyes was studied by UV-visible spectrophotometer (Hitachi U-2000) from the decrease in the respective λ_{max} values.

The colour intensity of the dyes was measured in terms of absorbance. Decolourisation was determined by using the calibration curve.

$$\text{Decolourisation (\%)} = \frac{C_0 - C}{C} \times 100 \tag{3}$$

where C_0, initial concentration of the dye; C, concentration of dye at time 't'.

2.5 Photodegradation studies

The photocatalytic degradation of dyes was evaluated in terms of total organic carbon (TOC). The total organic carbon of the pollutant was determined by using a TOC analyser (Shimadzu TOC V series).

2.6 Textile effluent study

To check the photocatalytic activity of M/TiO_2 catalyst for the photocatalytic degradation of actual textile effluent, the samples were collected from M/s Ramkay & Co, Erode, Tamil Nadu, India, filtered and diluted. Since the effluent contains number of unknown dyes it gave a broad peak in the wavelength range of 500–600 nm. Since the absorbance is additive, the effluent simply shows a broad peak in this range. To the diluted effluent (250 mL), 6 g of M/TiO_2 (M = Ag, Au and Pt) catalyst was added and the photocatalytic study was performed in the photocatalytic reactor.

2.6.1 COD experiment

Chemical oxygen demand (COD) of the samples collected at different intervals was determined by dichromate method in the presence of a catalyst Ag_2SO_4 [10, 11].

3. Results and discussion

3.1 Characterisation of synthesised TiO_2 and M/TiO_2 photocatalysts

3.1.1 UV-visible spectrophotometry

The absorption spectra of synthesised TiO_2 (gel form) was recorded using UV-visible spectrophotometer and given in **Figure 1**. An absorption maxima observed around 320 nm is due to the charge transfer from the VB of 2p orbitals of the oxide ions to the CB of $3d_{t2g}$ orbitals of the titania cations [12].

The optical absorption evolution spectra of solutions of noble metal precursors (2×10^{-3} M $AgNO_3$, 2×10^{-4} M $HAuCl_4$ and 1×10^{-5} M $[Pt(NH_3)_4]Cl_2$) in the presence of TiO_2 colloid during visible irradiation are shown in **Figure 1**.

This figure shows that on visible irradiation, appearance of new bands centred on 370 nm for Ag, 520 nm for Au and 502 nm for Pt were observed. The reason for

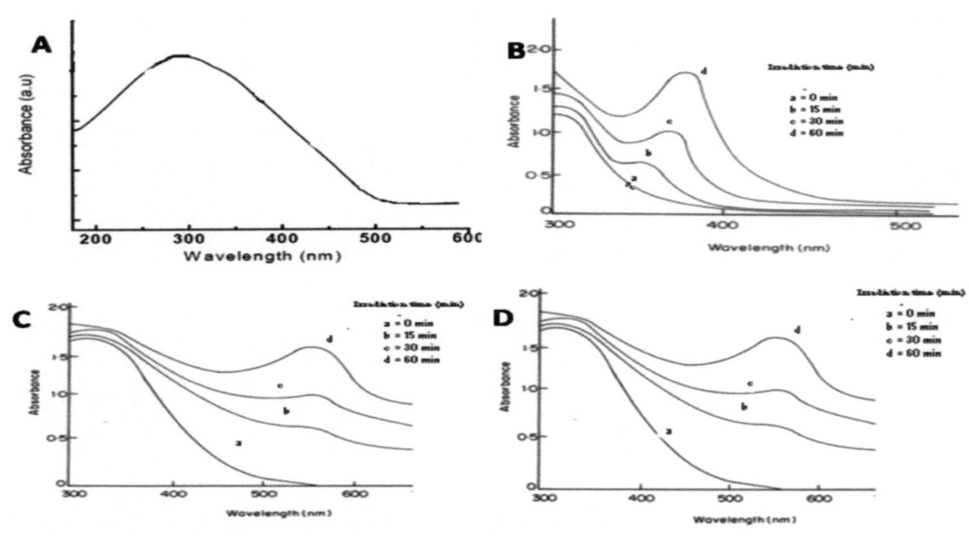

Figure 1.
(a) UV-visible spectrum of synthesised TiO_2 and optical absorption evolution spectra of (b) Ag/TiO_2 (c) Au/TiO_2 and (d) Pt/TiO_2 at different time intervals.

appearance of these peaks is due to the surface plasmon excitation of the respective metal colloids [13–15].

The band gap values of all the synthesised catalysts (TiO_2 and M/TiO_2) were calculated from the corresponding λ values obtained by extrapolating the rising portion of the spectrum to the × axis at zero absorbance and by using the following equation. The values were consistent with the values reported already [16].

$$E_g = hc/\lambda \tag{4}$$

where E_g, band gap energy; h, Planck's constant in eV (4.135×10^{-15} eV); c, velocity of light (3×10^8 m/s); λ, wavelength of corresponding M/TiO_2 catalysts 370 nm for Ag, 520 nm for Au and 502 nm for Pt.

The band gap values for the synthesised TiO_2 and metal loaded TiO_2 are given in **Table 2**. The band gap values were found to be lower for the metal loaded catalyst when compared to the synthesised TiO_2 catalyst.

Although lowering of band gap is not good for better catalytic activity under UV irradiation due to easy recombination, the presence of metals act as electrons traps and prevents the recombination process and also making the photocatalyst active in the visible range.

3.1.2 Atomic absorption spectrophotometric analysis

To determine the actual metal content of all the synthesised M/TiO_2 catalysts, they were subjected to AAS analysis after dissolving them. The metal content in each catalyst determined is given in **Table 2**. The results show that the experimentally determined metal content value is close to that of the theoretical value (1% w/w).

3.1.3 BET surface area measurements

The surface area of all the synthesised catalysts viz., TiO_2 (P-25 Degussa), TiO_2 and M/TiO_2 catalysts were determined and given in **Table 2. Table 2** clearly shows that the synthesised catalysts show higher surface than commercial TiO_2 (P-25 Degussa). It is to be noted that impregnation of noble metals (Pt, Pd and Au) over TiO_2 did not alter the surface area values significantly. A very small reduction in the surface area observed may be due to the blocking of fine capillaries present on TiO_2 surface by metal thin islands. These islands prevent the entry of the probe molecule (nitrogen gas) into the pores during BET measurement [17].

3.1.4 X-ray diffraction analysis

To obtain information regarding the phase formation and crystallite size, X-ray diffraction measurements were performed for the synthesised TiO_2 and M/TiO_2 (M = Ag, Au and Pt) photocatalysts and the XRD patterns are shown in **Figure 2**.

The synthesised TiO_2 has both anatase and rutile phases but not the brookite phase. Anatase and rutile phases are confirmed by the appearance of major peaks at $2\theta = 25.4$ and 48° respectively. The corresponding d (111) reflections of the noble metal atoms were found at $2\theta = 38.1$, 38.8 and 40° for Ag, Au and Pt, which confirms the impregnation of metal particles on TiO_2 lattice. From the X-ray diffraction patterns the average particle size of the synthesised TiO_2 and M/TiO_2 was calculated using Debye-Scherrer equation.

$$d = K\lambda/\beta \cos\theta \tag{5}$$

(A)

Catalyst	Band gap (eV)
TiO_2(P-25 Degussa)	3.2
Synthesised TiO_2	3.8
Ag/TiO_2	3.3
Au/TiO_2	2.3
Pt/TiO_2	2.4

(B)

Catalyst	Noble metal content (Wt. %)
Ag/TiO_2	0.97
Au/TiO_2	0.96
Pt/TiO_2	0.98

(C)

Catalysts	BET surface area (m^2g^{-1})
TiO_2(P-25 Degussa)	50.2
Synthesised TiO_2	90.5
Ag/TiO_2	87.3
Au/TiO_2	88.1
Pt/TiO_2	87.8

(D)

Catalyst	Particle Size (nm)
Synthesised TiO_2	24
Ag/TiO_2	11
Au/TiO_2	10
Pt/TiO_2	19

Table 2.
Physicochemical characteristics of bare and M/TiO_2 catalysts.

where d, diameter of the metal particle; λ, wavelength of the X-ray radiation (λ = 0.15418 nm), K = 0.98 (constant); θ, characteristic X-ray diffraction peak; B = full width at half maximum in radians.

The average particle diameter of the synthesised TiO_2 and M/TiO_2 was found to be in the nanometre range as shown in **Table 2**.

3.1.5 Transmission electron microscopy

TEM images and particle size histograms of synthesised TiO_2, Ag, Au and Pt deposited TiO_2 catalysts are shown in **Figure 3**. Almost uniform dispersion was obtained

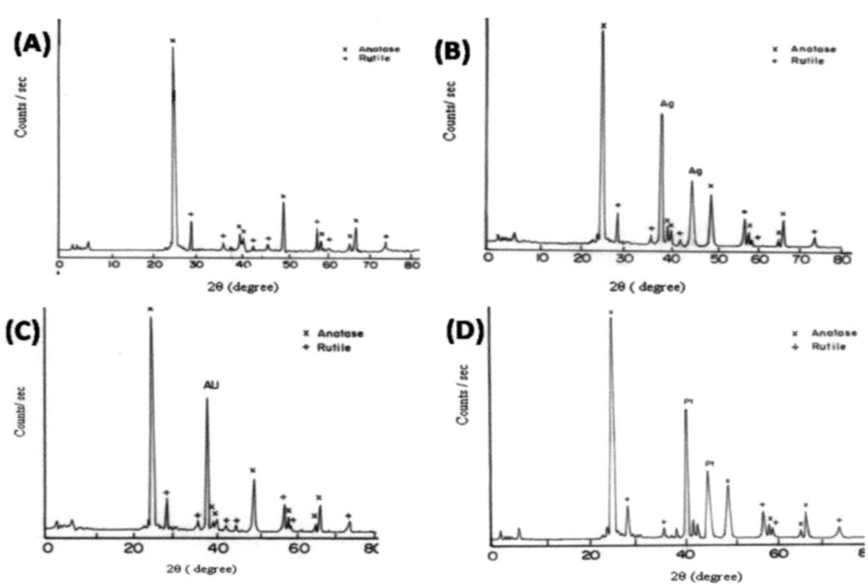

Figure 2.
X-ray diffraction patterns of (A). synthesised TiO_2, (B). Ag/TiO_2, (C) Au/TiO_2 and (D). Pt/TiO_2.

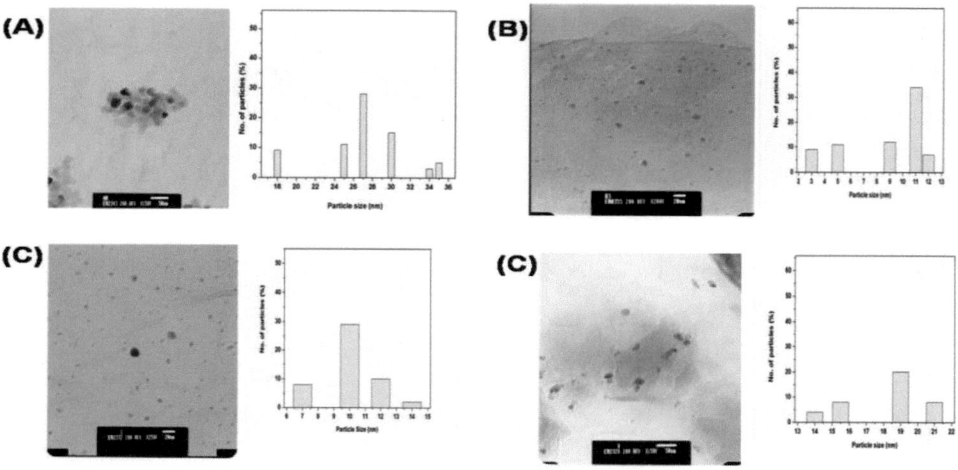

Figure 3.
Transmission electron micrograph and particle size histograms of (A) synthesised TiO_2, (B) Ag/TiO_2, (C)Au/TiO_2 and (D) Pt/TiO_2.

with all the catalysts. The average size of the particles was calculated by averaging particle sizes of 20–30 particles. The plot of particle size distribution is also shown.

The average particle size of TiO_2, Ag/TiO_2, Au/TiO_2 and Pt/TiO_2 catalyst was found to be 27 nm, 11 nm, 10 nm and 19 nm respectively.

3.2 Photocatalytic studies on TiO_2 (P-25 DEGUSSA)

3.2.1 Effect of initial concentration of dye

The effect of initial concentration of dye on the rate of decolourisation was studied by taking 250 mL of dye solutions and varying the concentrations between 8×10^{-5} and 1.3×10^{-4} M for Tartrazine (TAZ), 8×10^{-6} and 1.6×10^{-5} M for RY-17 and 6×10^{-6} and 1.6×10^{-5} for RB-5 with constant catalyst weight (1.5 g). The irradiation was carried out for 6 h by using 125 W low pressure mercury arc lamp (wave length 254 nm) and 85 W tungsten lamp (wave length 365 nm) as UV and visible light sources, respectively. **Figure 4** shows that percentage decolourisation decreases as the initial concentration of the dye increases under both UV and visible light illumination.

Similar results in the photocatalytic degradation of phenol were reported in the literature [18, 19]. The decolourisation rate relates to the probability of formation of hydroxyl radicals ($OH^{\cdot}$) on the catalyst surface and the probability of hydroxyl radicals reacting with dye molecules. Hence the rate constant depends on the probabilities of the formation of these two. Hence k' can be expressed as

$$k' = k_0 \times P_{OH} \times P_{dye\ molecules} \tag{6}$$

where k' is the overall rate constant and k_0 is the reaction rate constant. $P_{OH}^{\cdot}$ and $P_{dye\ molecules}$ refer to the probabilities of generation of $OH^{\cdot}$ radicals and $OH^{\cdot}$ radicals reacting with dye molecules. We all know that the reaction rate constant k_0 is independent of initial dye concentration but $P_{OH}^{\cdot}$ and $P_{dye\ molecules}$ will be affected by the dye concentration. Literature suggests that photocatalytic degradation of aromatic compounds mainly occurs by hydroxyl radicals [20]. The rate determining step of the reaction may be the formation of $OH^{\cdot}$ radicals that are formed through the reaction of holes with adsorbed OH^- and H_2O [21, 22].

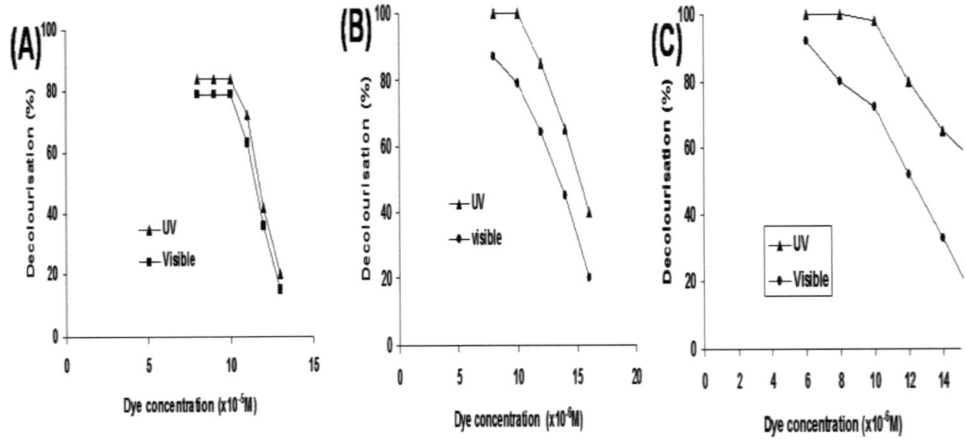

Figure 4.
Effect of initial concentration of (A)TAZ, (B) RY-17 and (C) RB-5 dyes on the % decolourisation under UV and visible irradiations. (reaction conditions: Weight of catalyst (P-25 Degussa) = 1.5 g, volume of dye solution = 250 mL, irradiation time = 6 h (UV): 6 h (visible) and pH = neutral).

As the dye concentration increases, the available hole sites may be occupied by dye ions which are generated from the sodium salt of the dye molecules,

$$[\text{Dye} - \text{Na}] \rightarrow [\text{Dye}]^{-} + \text{Na}^{+} \quad (7)$$

Since there are only a few active sites available for the generation of $OH^{\cdot}$ radicals the generation of $OH^{\cdot}$ will be reduced.

It is concluded that as the initial dye concentration increases, the catalyst surface needs to generate more amount of $OH^{\cdot}$ radicals and other oxidising species. But, illumination time and required amount of the catalyst are constant, so that $OH^{\cdot}$ radicals and other oxidising species formed on the TiO_2 surface are also constant. So the relative number of $OH^{\cdot}$ attacking the dye molecules decreases with increasing dye concentration [23]. Moreover, as the initial concentration of the dye increases, the path length of photons entering the solution decreases and at low dye concentration the reverse effect is observed [24, 25]. Also, at higher concentration, degradation decreases at sufficiently long distances from the light source or reaction zone due to retardation in the penetration of light. Hence, under both UV and visible light sources the rate of degradation decreases considerably with increase in dye concentration above the optimal concentration.

3.2.2 Effect of catalyst weight

A series of experiments were carried out to assess the optimum weight of the catalyst by varying the amount of TiO_2 (P-25 Degussa) from 0 to 3 g for the decolourisation of 250 mL of dye solution (TAZ = 1×10^{-4} M, RY-17 and RB-5 = 1×10^{-5} M) at the irradiation time of 6 h at neutral pH. The effects of catalyst weight on the percentage decolourisation of all the dyes are shown in **Figure 5**.

In the absence of the catalyst no decolourisation occurred. From the figure it is also evident that the rate of photodecolourisation increased linearly with the weight of catalyst up to 1.5 g irrespective of light sources. On increasing the catalyst weight further, the percentage decolourisation decreased. This decrease in decolourisation on increasing the catalyst weight may be due to the formation of turbidity (Shadowing effect). Similar observation was also reported [26]. Optimum catalyst

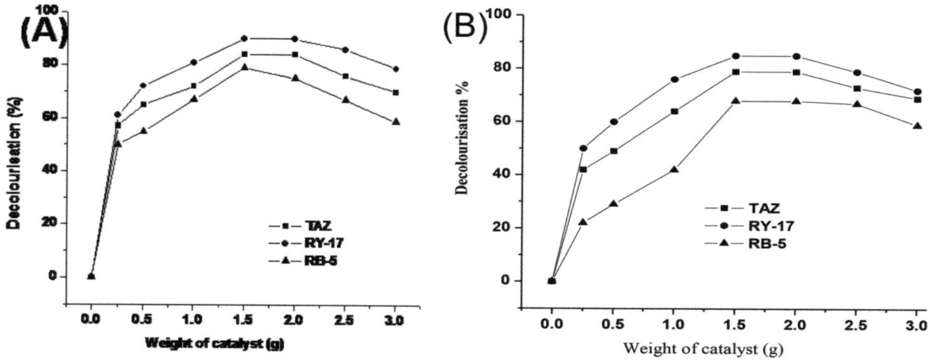

Figure 5.
Effect of weight of catalyst on the decolourisation of dyes under (A) UV and (B) visible irradiations (reaction conditions: Dye concentration: TAZ = 1 × 10^{-4} M, RY-17 = 1 × 10^{-5} M, RB-5 = 1 × 10^{-5} M, volume of dye solution = 250 mL, irradiation time = 6 h and pH = 7).

loading is always essential and the higher amount of TiO_2 may not be useful both in view of aggregation as well as reduced irradiation field due to shadowing effect [27].

As the weight of the catalyst increased, the quantity of photons adsorbed increased and consequently the decolourisation rate increased [28], Hence, optimisation of the catalyst loading for a given dye concentration is an important parameter to avoid excess catalyst and to ensure total absorption of light either UV or visible light for efficient photodecolourisation.

3.2.3 Effect of pH

The effect of pH on decolourisation of dyes irradiated under UV and visible irradiation are given in **Figure 6**.

The percentage decolourisation of all the dyes was found to be maximum at the pH of 7 for UV and visible irradiations. On increasing the pH from acidic to neutral under UV and visible irradiations, the percentage decolourisation increased significantly. However, when the pH was increased beyond these values to the basic ranges the percentage decolourisation decreased drastically. The main reaction is presented by the hydroxyl radical attack on the dye anion. Hydroxyl radicals

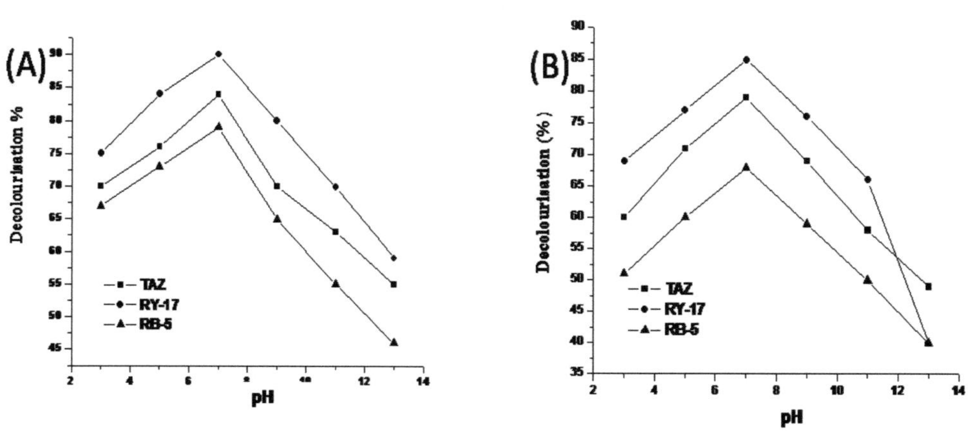

Figure 6.
Effect of pH on decolourisation of dyes under (A) UV and (B) visible irradiations (reaction conditions: Dye concentration: TAZ = 1 × 10^{-4} M, RY-17 = 1 × 10^{-5} M, RB-5 = 1 × 10^{-5} M, volume of dye solution =250 mL and irradiation time = 6 h).

which are considered as the predominant species at neutral or alkaline pH values are generated by oxidising more hydroxide ions [29]. At low pH values (pH < 5) the photodecolourisation of dyes is retarded under both UV and sunlight sources by the high concentration of the proton. Lack of availability of hydroxyl radicals in the pH range less than 5 leads to the decrease in decolourisation of dyes. In highly alkaline conditions (>pH 9) the % decolourisation of dyes decreased drastically due to the electrostatic columbic repulsion between the anionic dye surface (negatively charged) and the hydroxyl anions. Due to this repulsion, the dyes do not interact closely with the anions [30–33]. Thus it is deduced that the efficient condition for the maximum degradation of all the mentioned dyes is at neutral pH (pH = 7).

3.2.4 Photocatalytic degradation studies

The extent of degradation of dyes was followed by total organic carbon analyser (TOC). The experimental results revealed that as the irradiation time increased, some dye molecules may degrade into components of lower molecular fragments and they mineralise. TOC analysis was carried out for all the dye samples collected at different intervals of time and the results are shown in the **Figure 7**.

Since the decrease in the TOC content is the direct measure of degradation, TOC studies have been carried out to check whether the photocatalyst converts the harmful dye into harmless products. The studies revealed that, about 45% of TOC of dye samples was reduced under UV and 35% under visible irradiations. Similar type of observations was also made by [34, 35]. The degradation of dye involved the cleavage of —N = N— and sequential evolution of N_2 in the early stage of degradation [36]. The fate of nitrogen containing compounds in the photodegradation was also explained by [37].

3.2.5 Recycling of TiO_2

In order to test the recyclability of TiO_2 catalyst, the experiment was repeated with 250 mL of dye solutions (TAZ =1 × 10^{-4} M, RY-17 and RB-5 = 1 × 10^{-5} M) with the used catalyst. The photocatalytic experiments were carried out with 1.5 g of TiO_2 (P-25 Degussa) for 5 h. At the end of reaction, the dye samples were filtered and the catalyst was filtered off. There was not much difference between the weight of the recovered catalyst with its initial weight (<1%). Repeated experiments were carried out for another 5 h by taking 250 mL of fresh dye solution. The pseudo first order rate constant values were calculated for each cycle as mentioned before and the values are given in **Table 3**.

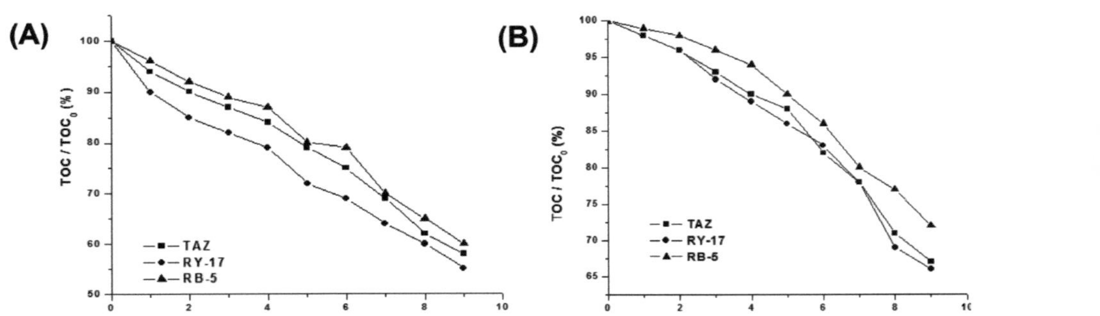

Figure 7.
Mineralisation of dyes using TiO_2 (P-25 Degussa) under (A) UV and (B) irradiations (reaction conditions: Dye concentration: TAZ = 1 × 10^{-4} M, RY-17 = 1 × 10^{-5} M, RB-5 = 1 × 10^{-5} M, volume of dye solution = 250 mL, irradiation time = 9 h and pH = 7).

No. of cycles	Pseudo first order rate constant K (min^{-1})					
	Tartrazine		Reactive Yellow-17		Reactive Black-5	
	UV	Visible	UV	Visible	UV	Visible
Fresh	8.7×10^{-3}	7.0×10^{-3}	9.4×10^{-3}	8.3×10^{-3}	6.3×10^{-3}	5.4×10^{-3}
I-recycle	8.15×10^{-3}	6.9×10^{-3}	9.1×10^{-3}	8.1×10^{-3}	5.9×10^{-3}	5.2×10^{-3}
II-recycle	8.09×10^{-3}	6.2×10^{-3}	8.2×10^{-3}	7.9×10^{-3}	5.3×10^{-3}	4.9×10^{-3}

Table 3.
Effect of catalyst recyclability on the degradation of textile dyes.

The rate constant values were found to be almost the same for all the dyes under both UV and visible irradiations even after two cycles. Nevertheless at the end of II recycle, 91% decolourisation occurred in the illumination period of 5 h. This shows that the TiO_2 (P-25 Degussa) is a very good catalyst which does not undergo significant deactivation even after 2 cycles. A very slight decrease in the rate constant values obtained in the photocatalysis experiments may be due to the adsorption of dye on the active sites of the catalysts which could not be removed during filtering.

3.2.6 Effect of inhibitors

Large amount of sodium chloride is used in the dyeing process in the textile industries and hence it usually comes out in the dye effluent along with sectional water of textile mills. Sodium carbonate also plays an important role in fixing the dye on fabrics and in the fastness of colour and hence it is mainly used in the dye bath to adjust the pH of the bath. Alcohols are used for padding process during dyeing the fabrics. Alcohols such as ethanol are commonly used to quench hydroxyl radicals. Hence, it is important to study the influence of chloride ions, carbonate ions and ethanol on the photomineralisation efficiency of the catalyst. Irradiation experiments were carried out by taking 250 mL of different dyes of various concentration (TAZ =1×10^{-4} M, RY-17 and RB-5 = 1×10^{-5} M), 1.5 g of TiO_2 (P-25 Degussa) under neutral pH and with different concentrations of (0–2 g) for both sodium chloride and sodium carbonate and 0 to 1 mL of ethanol. The irradiation was carried out by using 125 W low pressure mercury arc lamp and 85 W tungsten lamp as UV and visible light sources, respectively. The results obtained under both UV-visible irradiations are shown in **Figure 8**.

The presence of chloride ion in the dye samples had a negative effect on decolourisation. The percentage decolourisation decreased drastically with increase in the concentration of sodium chloride. This decrease in the percentage decolourisation of dyes in the presence of chloride ion is due to the hole-scavenging properties of chloride ion (equations 7 and 8).

$$TiO_2 \rightarrow TiO_2(h^{+}{}_{vb}, e^{-}{}_{cb}) \quad (8)$$

$$Cl^{-} + h^{+}{}_{vb} \rightarrow Cl^{\cdot} \quad (9)$$

$$Cl^{\cdot} + Cl^{-} \rightarrow Cl_2^{\cdot -} \quad (10)$$

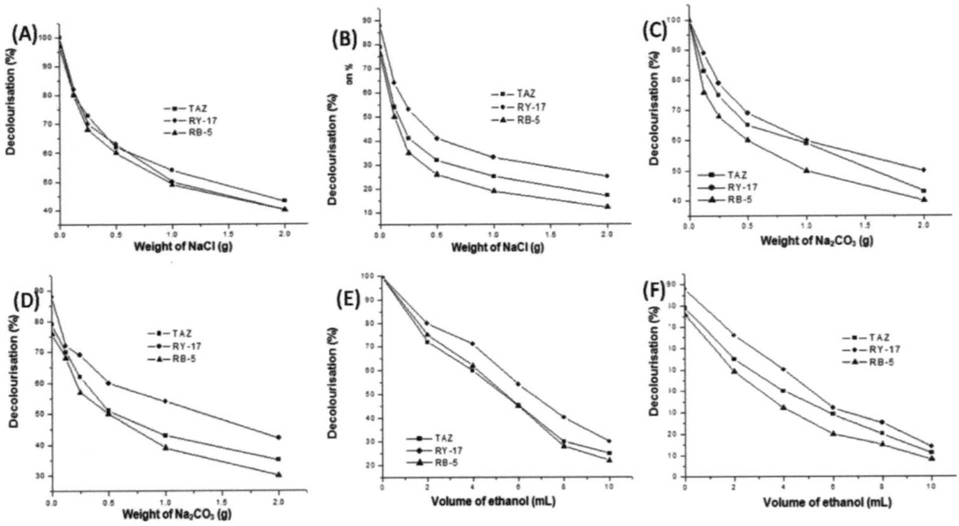

Figure 8.
Effect of (A) sodium chloride, (C) sodium carbonate and (E) ethanol on decolourisation of dyes under UV and (B, D, F) visible irradiations. (reaction conditions: Dye concentration: TAZ = 1 × 10^{-4} M, RY-17 = 1 × 10^{-5} M, RB-5 = 1 × 10^{-5} M, weight of catalyst (P-25 Degussa) = 1.5 g, volume of dye solution = 250 mL, pH = neutral and irradiation time = 6 or 6 ½ h).

This is a distinctive competitive inhibition reaction where the reaction of dye molecules with the holes has to compete with the chloride ions, which are not readily oxidisable. This inhibitive effect was also observed by [38]. However, the photocatalytic activity of the catalyst can be fully restored by washing the catalyst with pure water.

The decrease in the degradation of dyes in the presence of carbonate and bicarbonate ions is due to the hydroxyl scavenging property of carbonate ions according to the following Eqs. (11) and (12).

$$OH^{\cdot} + CO_3^{2-} \rightarrow OH^{-} + CO_3^{\cdot -} \quad (11)$$

$$OH^{\cdot} + HCO_3^{-} \rightarrow H_2O + CO_3^{\cdot -} \quad (12)$$

The presence of carbonate and bicarbonate ions reduces the concentration of hydroxyl radicals significantly and hence the percentage decolourisation of the dyes decreases [39].

When the volume of ethanol was increased the photodecolourisation efficiency gradually decreased. This shows that the dye degradation occurs via positive holes (h^{+}_{vb}) that are not involved in recombination with electrons during illumination process. Further it is evident that hydroxyl radicals are not the only species responsible for degradation and minor degradation also occurs through holes generated during illumination [40].

3.2.7 Effect of electron acceptors

The practical problem in using TiO_2 as a photocatalyst is the undesired electron/hole recombination. In the absence of proper electron acceptors or donors the electron/hole recombination is extremely efficient and represents the major energy wasting step which limits the achievable quantum yield. One strategy to inhibit electron/hole recombination is to add other electron acceptors such as hydrogen peroxide, persulphate ions to the reaction mixture.

Hydrogen peroxide is extensively used in commercial bleaching of textiles, especially cellulose fibres and wool. In order to study the influence of hydrogen peroxide in the decolourisation of dyes, experiments were performed by adding different quantities of hydrogen peroxide(30% v/v) (0–10 mL) and 0 to 2 g of potassium persulphate to 250 mL of different dyes of various concentration (TAZ = 2×10^{-4} M, RY-17 and RB-5 = 2×10^{-5} M), 1.5 g of TiO_2 (P-25 Degussa) under neutral pH. The irradiation was carried out by using 125 W low pressure mercury arc lamp and 85 W tungsten lamp as UV and visible light sources respectively.

The presence of hydrogen peroxide will prevent the (e^-/h^+) pair recombination and enhances the formation of hydroxyl radicals (Eq. 13).

$$TiO_{2(e-)} + H_2O_2 \rightarrow TiO_2 + OH^- + OH^{\cdot} \tag{13}$$

Since hydrogen peroxide is better than molecular oxygen as electron acceptor it leads to 100% decolourisation in comparison to the results obtained in the absence of H_2O_2. But, in the absence of TiO_2, and with only H_2O_2 the percentage decolourisation was found to be less under both the light sources as shown in **Figure 9**.

The percentage decolourisation increased on increasing the volume of H_2O_2 from 0 to 10 mL. However, percentage decolourisation decreased on increasing the volume of H_2O_2 beyond 10 mL. Since, H_2O_2 is a powerful hydroxyl scavenger, the percentage decolourisation decreased at the higher concentration as shown in Eqs. (14) and (15).

$$H_2O_2 + OH^{\cdot} \rightarrow HO_2^{\cdot} + H_2O \tag{14}$$

$$HO_2^{\cdot} + OH^{\cdot} \rightarrow H_2O + O_2 \tag{15}$$

The above decolourisation results carried out by both UV and visible sources show the necessity of choosing the proper dosage of hydrogen peroxide for maximum decolourisation. Thus, it is clear from the **Figure 9** that addition of H_2O_2 up to 10 mL concentration showed a beneficial effect on the photocatalytic decolourisation of the dyes [41].

The rate of photo-assisted decolourisation of dyes is significantly improved by the presence of persulphate ions.

The percentage decolourisation increased with increase in the amount of $K_2S_2O_8$ and 100% degradation was attained at highest concentration of $K_2S_2O_8$ for both UV and visible irradiations. The added persulphate ion acts as electron traps resulting in the formation of reactive radical intermediate $SO_4^{\cdot-}$ as shown in Eq. (16).

$$S_2O_8^{2-} + e^-_{cb} \rightarrow SO_4^{\cdot-} + SO_4^{2-} \tag{16}$$

The sulphate radical anion is a strong oxidant and removes electron from neutral molecules like water and generates hydroxyl radical.

$$SO_4^{\cdot-} + H_2O \rightarrow OH^{\cdot} + SO_4^{2-} + H^+ \tag{17}$$

The persulphate ions oxidise the dyes as well as act as electron scavenger. Due to their scavenging property they inhibit e^-/h^+ pair recombination at the semiconductor surface [42]. $S_2O_8^{2-}$ ions produce sulphate radical anion ($SO_4^{\cdot-}$) which in turn produce highly oxidative hydroxyl radicals. These hydroxyl radicals degrade the dyes at a faster rate.

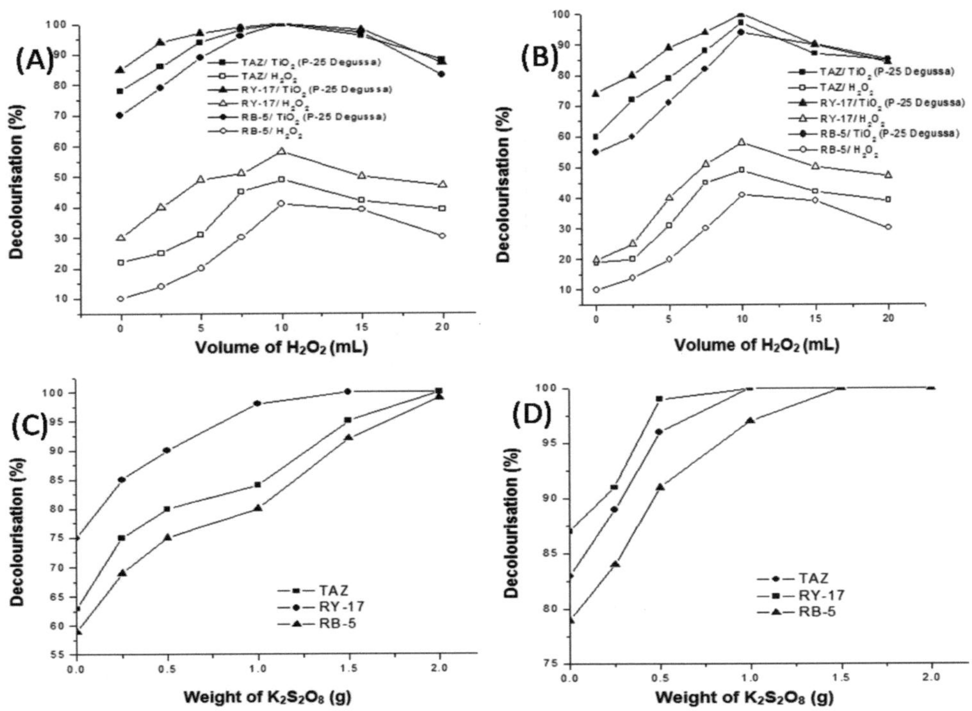

Figure 9.
Effect of (A) H_2O_2 and (B) $K_2S_2O_8$ on decolourisation of dyes under visible and (C and D) UV irradiations. (reaction conditions: Dye concentration: TAZ = 2×10^{-4} M, RY-17 = 2×10^{-5} M, RB-5 = 2×10^{-5} M, weight of catalyst (P-25 Degussa) = 1.5 g, volume of dye solution =250 mL, pH = neutral and irradiation time = 7 h).

3.3 Comparison of photocatalytic activities of synthesised TiO_2 with TiO_2 (P-25 Degussa)

The photocatalytic decolourisation of the dyes TAZ, RY-17 and RB-5 were carried out with the synthesised TiO_2 catalyst under the conditions optimised for the best decolourisation with TiO_2 (P-25 Degussa). The optimised conditions for the complete decolourisation of various dyes are given in **Table 4** and the time taken for complete decolourisation of the dyes over the synthesised titania is given in **Table 5**.

Among the three dyes, RY-17 reached complete decolourisation at a shorter irradiation time under both the light sources. RB-5 took longer irradiation time for complete decolourisation. This may be due to its chemical nature (i.e.) RB-5 being diazo dye. Diazo dyes are less degradable than mono azo dye during ozonation [43]. Similar trend of decolourisation was observed with visible light irradiation.

Dye	Initial concentration [M]	pH	Time taken for complete Decolourisation (h)		Weight of catalyst (g) TiO_2 (P-25 Degussa)
			UV irradiation	Visible irradiation	
TAZ	1×10^{-4}	7	5 ½	6 ½	1.5
RY-17	1×10^{-5}	7	5	6	1.5
RB-5	1×10^{-5}	7	6	7	1.5

Table 4.
The optimised conditions for the complete decolourisation of different dyes using TiO_2 (P-25 Degussa).

Catalyst	Time taken for complete decolourisation (h)					
	TAZ		RY-17		RB-5	
	UV	Visible	UV	Visible	UV	Visible
TiO_2 (P-25 Degussa)	5½	6½	5	6	6	7
Synthesised TiO_2	4	3	3½	4	4½	5

Table 5.
Time taken for complete decolourisation (100%) of various dyes using TiO_2 (P-25 Degussa) and synthesised TiO_2.

However, the times at which complete decolourisation took place were different and higher for visible light when compared with UV.

The higher efficiency of the synthesised catalyst when compared to P-25 Degussa may be due to the larger surface area of the synthesised TiO_2 (90.5 m^2/g) when compared with TiO_2 (P-25 Degussa) (50 m^2/g). Similar type of results was also observed by [44]. Because of the increase in the surface area of synthesised TiO_2 there may be increase in the adsorption of the three dyes.

From the above discussions, it is clear that both degradation and decolourisation of the dyes are highly effective with synthesised TiO_2 compared to TiO_2 (P-25 Degussa) under both the light sources. To further the photocatalytic activity further in the visible region, the modification in the synthesised catalyst has to be carried out by noble metal deposition. Noble metals such as Ag, Au and Pt have been deposited on the synthesised catalyst and they were evaluated for their photocatalytic activity.

3.4 Photodecolourisation studies of metal impregnated titania photocatalysts

To evaluate the photocatalytic activity of noble metal deposited titania catalyst, the decolourisation of dyes (TAZ, RY-17 and RB-5) was carried out under visible and UV light irradiations. The reaction was carried out with 250 mL of dyes of various concentration (TAZ =1×10^{-4} M, RY-17 and RB-5 = 1×10^{-5} M) and 1.5 g of M/TiO_2 catalyst at neutral pH. The irradiation was carried out by using 125 W low pressure mercury arc lamp (wave length 254 nm) and 85 W tungsten lamp (wave length 365 nm) as UV and visible light sources respectively. The results obtained in the decolourisation of various dyes with M/TiO_2 under UV and visible irradiations are shown in **Figures 10** and **11**.

From the above photocatalytic decolourisation studies the following observations were made: Photocatalytic efficiency of the different catalysts for the decolourisation of different dyes were found to be in the following order:

$$Au/TiO_2 > Ag/TiO_2 \sim Pt/TiO_2 > \text{synthesised } TiO_2 > TiO_2 \text{ (P - 25 Degussa)}$$

The time at which the complete decolourisation occurred for the different catalysts is given in **Table 6**.

- As far as the decolourisation of the dyes are concerned, all the catalysts found to decolourise RY-17 to the maximum extent followed by TAZ and RB-5.

The percentage decolourisation of different dyes obtained with the different catalysts were determined at the time at which the Au/TiO_2 catalyst showed 100% decolourisation of the respective dyes (TAZ, RY-17 and RB-5) using different catalysts are given in **Table 7**.

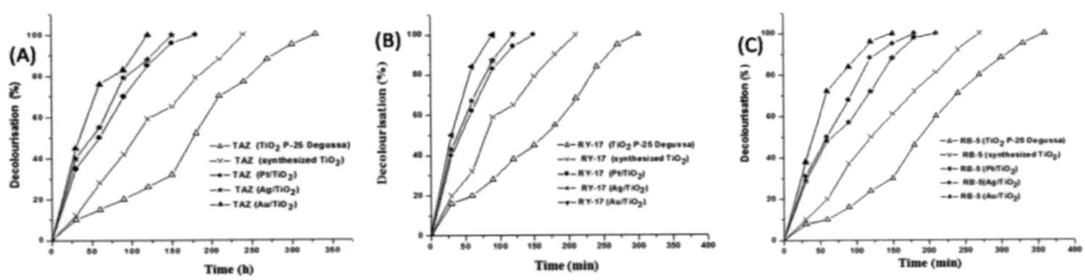

Figure 10.
Decolourisation of (A)TAZ, (B) RY-17 and (C) RB-5 using different TiO_2 catalysts under UV irradiation. (reaction conditions: Dye concentration: TAZ = 1 × 10^{-4} M, both RY-17 & RB-5 = 1 × 10^{-5} M, weight of catalyst = 1.5 g, volume of dye solution = 250 mL and pH = neutral).

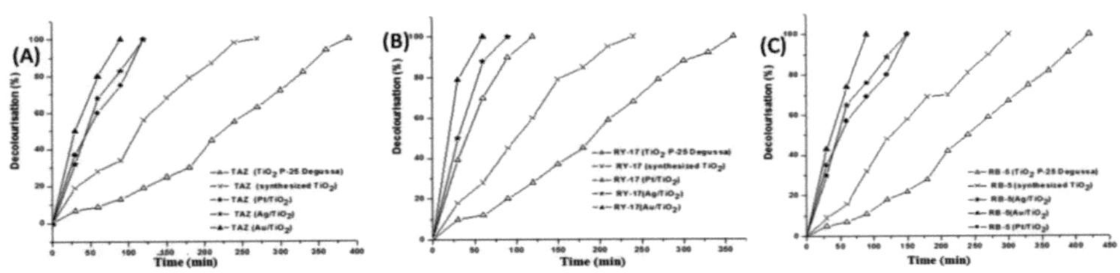

Figure 11.
Decolourisation of (A)TAZ, (B) RY-17 and (C) RB-5 using different TiO_2 catalysts under visible irradiation. (reaction conditions: Dye concentration: TAZ = 1 × 10^{-4} M, both RY-17 & RB-5 = 1 × 10^{-5} M, weight of catalyst = 1.5 g, volume of dye solution = 250 mL and pH = neutral).

Among all the catalysts, M/TiO_2 catalysts show the higher photocatalytic activity towards the decolourisation of all the three dyes. The higher catalytic activity of M/TiO_2 catalysts may be due to the smaller band gap values when compared to TiO_2 (P-25 Degussa) and the ability of the noble metals trap the electrons, thus preventing the recombination. The rate constant values obtained with different M/TiO_2 catalysts in the photocatalytic decolourisation of all the three dyes were found to be highest for Au/TiO_2 followed by Ag/TiO_2 and Pt/TiO_2. The higher photocatalytic activity of the Au/TiO_2 catalyst when compared to the other metals is due to the lowered band gap values [45, 46], which leads to the effective absorption of visible light by the catalyst resulting in higher decolourisation.

3.5 Photodegradation studies

The photodegradation studies was carried out with 250 mL of dyes of various concentration (TAZ = 1 × 10^{-4} M, RY-17 and RB-5 = 1 × 10^{-5} M) and 1.5 g of M/TiO_2 catalyst at neutral pH. The irradiation was carried out by using 125 W low pressure mercury arc lamp (wave length 254 nm) and 85 W tungsten lamp (wave length 365 nm) as UV and visible light sources respectively. The degradation was monitored by a TOC analyser and the results are shown in **Figure 12**.

All the above photodegradation experiments show that like photodecolourisation that photodegradation efficiency of all the synthesised catalyst follow similar trend as mentioned below:

- Au/TiO_2 > Ag/TiO_2 ≈ Pt/TiO_2 > Synthesised TiO_2 > TiO_2 (P-25 Degussa)
- As far as the dyes are concerned, all the catalysts found to degrade RY-17 to the maximum extent followed by TAZ and RB-5.

Dye	Time taken for complete Decolourisation (h)									
	TiO_2 (P-25 Degussa)		Synthesised TiO_2		Pt/TiO_2		Ag/TiO_2		Au/TiO_2	
	UV	Visible	UV	Visible	UV	Visible	UV	Visible	UV	Visible
TAZ	5 ½	6 ½	4	4 ½	3	2	2 ½	2	2	1 ½
RY-17	5	6	3 ½	4	2 ½	2	2	1 ½	1 ½	1
RB-5	6	7	4 ½	5	3 ½	2 ½	3	2 ½	2 ½	1 ½

Table 6.
Time taken for the 100% decolourisation of the dyes (TAZ, RY-17 and RB-5) using different catalysts.

Dye	Decolourisation (%)									
	TiO_2 (P-25 Degussa)		Synthesised TiO_2		Pt/TiO_2		Ag/TiO_2		Au/TiO_2	
	UV	Visible	UV	Visible	UV	Visible	UV	Visible	UV	Visible
TAZ	26	16	79	75	79	75	85	83	100	100
RY-17	28	17	83	80	83	80	90	88	100	100
RB-5	24	11	70	72	70	72	88	76	100	100

Table 7.
Percentage decolourisation of the different dyes (TAZ, RY-17 and RB-5) using different catalysts.

The degradation percentage of the dyes (TAZ, RY-17 and RB-5) using different catalysts at the time period of 8 h are given in **Table 8**.

The photocatalytic studies carried out under UV and visible irradiations leads to the following observations:

- The photocatalytic activity of the synthesised catalyst in the decolourisation and degradation of all the dyes was found to be better than TiO_2 (P-25 Degussa) under both UV and visible irradiations.

- Higher percentage of decolourisation and degradation occurred for both TiO_2 (P-25 Degussa) and the synthesised catalyst under UV irradiation when compared to visible irradiation.

- The photocatalytic activities of M/TiO_2 catalysts were found to be higher under both UV and visible irradiations than the synthesised TiO_2.

- The photocatalytic efficiency of the M/TiO_2 was remarkable even under visible light irradiation when compared to the UV irradiation.

The higher activity of M/TiO_2 catalysts under visible region is explained below:

- The activity of noble metals may be attributed to the electronic interaction between the impregnated noble metal and TiO_2. Due to this interaction, Schottky barriers are formed leading to electron hole pair [45].

- The metal deposits traps the electrons and quickly transfers the electrons to oxygen present in the solution to form the highly oxidative $O_2^{\cdot -}$ radical [47].

- The trapping rate of electrons by noble metals is faster rate when compared to the electron transfer from the semiconductor to O_2 [48].

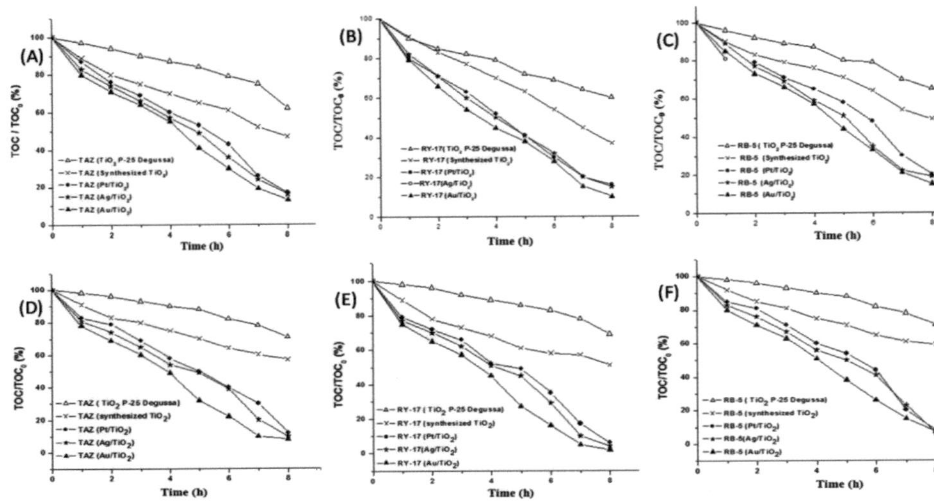

Figure 12.
Degradation of (A) TAZ, (B) RY-17 and (C) RB-5 using different TiO_2 catalysts under UV (D,E and F) visible irradiations. (reaction conditions: Dye concentration: TAZ = 1×10^{-4} M, both RY-17 & RB-5 = 1×10^{-5} M, weight of catalyst = 1.5 g, volume of dye solution = 250 mL and pH = neutral).

Dyes	Degradation (%)									
	TiO_2 (P-25 Degussa)		Synthesised TiO_2		Pt/TiO_2		Ag/TiO_2		Au/TiO_2	
	UV	Visible	UV	Visible	UV	Visible	UV	Visible	UV	Visible
TAZ	38	29	53	43	83	88	84	90	84	90
RY-17	40	31	63	49	84	96	85	96	85	96
RB-5	35	23	51	41	80	82	81	93	81	94

Table 8.
Percentage degradation of the dyes (TAZ, RY-17 and RB-5) using different catalysts.

3.6 Effluent studies

3.6.1 Photocatalytic activity of M/TiO_2 with the actual effluent sample

The photocatalytic activity of the synthesised M/TiO_2 catalysts (M = Ag, Au and Pt) was tested with the real effluent samples collected from a textile industry (M/s Ramkay, Erode, Tamil Nadu, India).

The industrial effluent (1 L) sample was first filtered to remove any insoluble matter (carbonates, hydroxides, etc.). As the initial pH was in weakly acidic, it was increased to 7 by adding dil. NH_4OH and the pH adjusted effluent was then subjected to photocatalytic treatment by adding 6 g of the M/TiO_2 (M = Ag, Au and Pt) catalyst. Before the lamp was switched on, the effluent was equilibrated for few minutes with continuous stirring. The effluent was then irradiated using visible light source. Aliquots of the sample were withdrawn at regular time intervals and the extent of decolourisation was monitored using UV-Vis spectrophotometer (Hitachi U 2000). The degradation of the dyes was also determined by using total

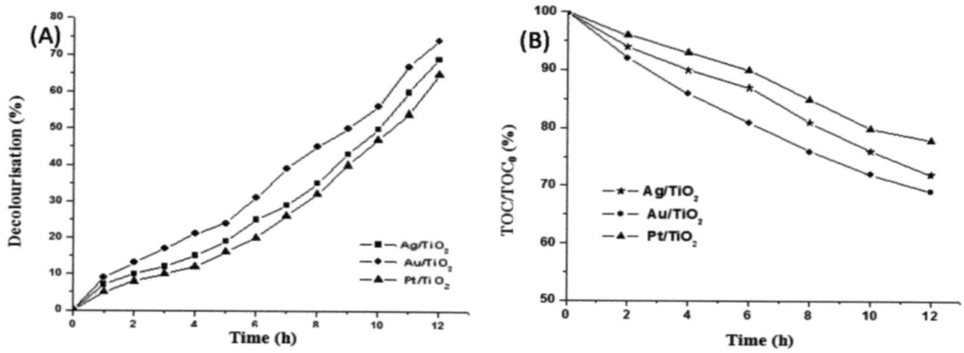

Figure 13.
(A) Decolourisation and (B) degradation of textile effluent using M/TiO$_2$ catalysts. (reaction conditions: weight of catalyst = 6 g, volume of effluent = 1 L and pH = neutral).

organic carbon analyser (TOC). The decolourisation and degradation results obtained are shown in **Figure 13**.

The effluent studies show that, all the M/TiO$_2$ catalysts decolourise and degrade the effluent samples significantly. However, these catalysts decolourise the effluent to the extent of only 65–75% and degrade to the extent of 25–30% that too after longer time of irradiation (12 h). The lesser percentages of decolourisation and degradation may be due to the presence of infinite number of dyes, salts etc. in the effluent.

4. Conclusions

The optimised reactions parameters for the degradation of TAT, RY-17 and RB-5 were found to be:

The initial dye concentration (TAZ = 1×10^{-4} M and for both RY-17 and RB-5 = 1×10^{-5} M), catalyst weight (1.5 g/L), pH (7) and the time taken for complete decolourisation (TAZ: 5 ½ h (UV) 6 ½ h (visible), RY-17: 5 h (UV) 6 h (visible), RB-5: 6 h (UV) 7 h (visible). Decolourisation and degradation were comparatively faster under UV than under visible. Inhibitors such as sodium chloride, sodium carbonate and ethanol decreased the degradation rate and electron acceptors such as H_2O_2 and potassium per sulphate showed beneficial effect. As far as the photocatalytic efficiency of the catalysts are concerned, the following order was observed under both UV and visible irradiations,

$$Au/TiO_2 > Ag/TiO_2 \sim Pt/TiO_2 > \text{Synthesised } TiO_2 > TiO_2\ (P-25\ \text{Degussa})$$

Among the noble metals, Au/TiO$_2$ catalyst was found to be the most active catalyst. This may be attributed to the small band gap value of Au/TiO$_2$ leading to better visible light absorption.

All the M/TiO$_2$ catalysts were found to be more active towards the decolourisation of all the dyes even under visible irradiation. The enhanced photocatalytic activity of the M/TiO$_2$ under visible light irradiation may be due to the presence of impregnated metals, which act as electron traps and by this way they prevent electron hole recombination. The synthesised M/TiO$_2$ catalysts were found to be very active towards the decolourisation and degradation of textile dye effluent collected from an industry as well.

Acknowledgements

The authors sincerely thank the funding agencies such as MOEF, and DRDO for funding research schemes to Catalysis Laboratory, Applied Sc. And Tech, ACTech, Anna University, Chennai, India.

Conflict of interest

The authors have no conflict of interest.

Author details

Sivakumar Thiripuranthagan* and Valentine Rupa
Department of Applied Science and Technology, Anna University, Chennai, Tamil Nadu, India

*Address all correspondence to: tssivakumar@yahoo.com

References

[1] Manikavasagam N. Treatment of Textile Processing Effluents. Coimbatore, India: Sakthi Publications; 1995

[2] Hachem C, Bocquillon F, Zahraa O, Bouchy M. Decolourization of textile industry wastewater by the photocatalytic degradation process. Dyes and Pigments. 2001;**49**:117-125. DOI: 10.1016/S0143-7208(01)00014-6

[3] Bickley RI, Carreno TG, Lees JS, Palmisano L, Tilley RJD. A spectral investigation of titanium dioxide photocatalysts. Journal of Solid State Chemistry. 1992;**92**:178-190. DOI: 10.1016/0022-4596(91)90255-G

[4] Linsebigler AL, Lu G, Yates JT. Photocatalysis on TiO_2 surfaces: Principles, mechanisms, and selected results. Chemical Reviews. 1995;**5**: 735-758. DOI: 10.1021/cr00035a013

[5] Howe RF. Recent developments in photocatalysis. Developments in Chemical Engineering and Mineral Processing. 1998;**6**(1):55-84. DOI: 10.1002/apj.5500060105

[6] Renault NJ, Pichat P, Foissy A, Mercier R. Study of the effect of deposited platinum particles on the surface charge of titania aqueous suspensions by potentiometry, electrophoresis, and labeled-ion adsorption. The Journal of Physical Chemistry. 1986;**90**:2733-2738. DOI: 10.1021/j100403a035

[7] Hadjiivanov KI, Klissurski DG. Surface chemistry of titania (anatase) and titania-supported catalysis. Chemical Society Reviews. 1996:61-69. DOI: 10.1039/CS9962500061

[8] Sanchez E, Lopez T. Effect of the preparation method on the band gap of titania and platinum-titania sol-gel materials. Materials Letters. 1995;**25**:271-275. DOI: 10.1016/0167-577X(95)00190-5

[9] Yang JC, Kim YC, Shul YG, Shin CH, Lee TK. Characterization of photoreduced Pt/TiO_2 and decomposition of dichloroacetic acid over photoreduced Pt/TiO_2 catalysts. Applied Surface Science. 1997;**121**:525-529. DOI: 10.1016/S0169-4332(97)00359-0

[10] Moore WA, Kroner RC, Ruchhoft CC. Dichromate reflux method for determination of oxygen consumed. Analytical Chemistry. 1949;**21**(8): 953-957. DOI: 10.1021/ac60032a020

[11] Dobbs RA, Williams RT. Elimination of chloride interference in the chemical oxygen demand test. Analytical Chemistry. 1963;**35**(8):1064-1067. DOI: 10.1021/ac60201a043

[12] Lee BY, Park SH, Kang M, Lee SC, Choung SJ. Preparation of Al/TiO_2 nanometer photocatalyst film and the effect of H_2O addition on photocatalytic performance for benzene removal. Applied Catalysis A: General. 2003;**253**:371-380. DOI: 10.1016/S0926-860X(03)00542-8

[13] Wang C, Liu C, Zheng X, Chen J, Shen T. The surface chemistry of hybrid nanometer sized particles 1. Photodeposition of gold on ultrafine TiO_2 particles. Colloids and Surfaces, A: Physicochemical and Engineering Aspects. 1998;**131**:271-280

[14] Hahakura S, Isoda S, Ogawa T, Moriguchi S, Kobayashi T. Formation of ultrafine platinum particles in an aqueous solution with a surfactant. Journal of Crystal Growth. 2002;**237-239**(3):1942-1945

[15] Sobana N, Muruganandham M, Swaminathan M. Nano Ag particle doped TiO_2 for efficient photodegradation of direct azo dyes. Journal of Molecular Catalysis.

2006;**258**:124-132. DOI: 10.1016/j.molcata.2006.05.013

[16] Reddy EP, Davydov L, Smirniots P. Characterization of titania loaded V-, Fe-, and Cr- incorporated MCM-41 by XRD, TPR, UV-Vis, Raman and XPS techniques. The Journal of Physical Chemistry. B. 2002;**106**:3394-3401. DOI: 10.1021/jp013898

[17] Sakthivel S, Shankar MV, Palanichamy M, Arabindoo B, Bahnemann DW, Murugesan V. Enhancement of photocatalytic activity by metal deposition: Characterisation and photonic efficiency of Pt, Au and Pd deposited on TiO_2 catalyst. Water Research. 2004;**38**(13):3001-3008. DOI: 10.1016/j.watres.2004.04.046

[18] Wei TY, Wan CC. Heterogeneous photocatalytic oxidation of phenol with titanium dioxide powders. Industrial and Engineering Chemistry Research. 1991;**30**:1293-1300. DOI: 10.1021/ie00054a033

[19] Wang KS, Chen HY, Huang LC, Su YC, Chang SH. Degradation of reactive black 5 using combined electrochemical degradation-solar-light/immobilized TiO_2 film process and toxicity evaluation. Chemosphere. 2008;**72**(2):299-305

[20] Matthews RW. Hydroxylation reactions induced by near-ultraviolet photolysis of aqueous titanium dioxide suspensions. Journal of the Chemical Society, Faraday Transactions. 1984;**80**:457-471. DOI: 10.1039/F19848000457

[21] Okamoto KI, Yamamoto Y, Tanaka H, Tanaka M. Heterogeneous photocatalytic decomposition of phenol over TiO_2, powder. Bulletin of the Chemical Society of Japan. 1985;**58**:2015-2022

[22] Okamoto KI, Yamamoto Y, Tanaka H, Tanaka M. Kinectics of heterogeneous photocatalytic decomposition of phenol over anatase TiO_2 powder. Bulletin of the Chemical Society of Japan. 1985b;**58**:2023-2028

[23] Rideh L, Wehrer A, Ronze D, Zoulalian A. Photocatalytic degradation of 2-chlorophenol in TiO_2 aqueous suspension : Modeling of reaction rate. Industrial and Engineering Chemistry Research. 1997;**36**:4712-4718

[24] Matthews RW. Photooxidation and adsorption of methylene blue on thin films of near-ultraviolet-illuminated TiO_2. Journal of the Chemical Society, Faraday Transactions. 1989;**1, 85**(6):1291-1302. DOI: 10.1039/F19898501291

[25] Davis RJ, Gainer JL, Neal GO, Wenwu I. Photocatalytic decolourisation of wastewater dyes. Water Environment Research. 1994;**66**:50-53. DOI: 10.2175/WER.66.1.8

[26] Rao KVS, Lavédrine B, Boule P. Influence of metallic species on TiO_2 for the photocatalytic degradation of dyes and dye intermediates. Journal of Photochemistry and Photobiology A: Chemistry. 2003;**154**(2-3):189-193. DOI: 10.1016/S1010-6030(02)00299-X

[27] Suri RPS, Liu J, Hand DW, Crittenden JC, Perram DL, Mullins ME. Heterogeneous photocatalytic oxidation of hazardous organic contaminants in water. Water Environment Research. 1993;**65**:665-673

[28] Chen JR, Xu XW, Lee AS, Yen TFA. Feasibility study of dechlorination of chloroform in water by ultrasound in the presence of hydrogen peroxide. Environmental Technology. 1990;**11**:829-836. DOI: 10.1080/09593339009384930

[29] Stumm W, Morgan JJ. Aquatic Chemistry. New York: John Wiley and Sons; 1981

[30] Hidaka H, Nohara K, Zhao J. Photo-oxidative degradation of the pesticide permethrin catalysed by irradiated TiO_2 semiconductor slurries in aqueous media. Journal of Photochemistry and Photobiology A: Chemistry. 1992a;**64**:247-253. DOI: 10.1016/1010-6030(92)85111-7

[31] Hidaka HJ, Zhao J, Pelizzetti E, Serpone N. Photodegradation of surfactants 8. Comparison of photocatalytic process between anionic sodium dodecyl benzene sulfonate and cationic benzyl dodecyl dimethyl ammonium chloride on the TiO_2 surface. The Journal of Physical Chemistry. 1992b;**96**:2226-2229

[32] Zhao J, Wu T, Wu K, Oikawa K, Hidaka H, Serpone N. Photoassisted degradation of dye pollutants. Degradation of the cationic dye Rhodamine B in aqueous anionic surfactant/TiO_2 dispersions under visible light irradiation: Evidence for the need for substrate adsorption TiO_2 particles. Journal of Environmental Science and Technology. 1998;**32**: 2394-2401. DOI: 10.1021/es9707926

[33] Kormann C, Bahnemann DW, Hoffmann MR. Photolysis of chloroform and other organic molecules in aqueous TiO_2 suspensions. Environmental Science & Technology. 1991;**25**:494-500. DOI: 10.1021/es00015a018

[34] Mansila HD, Villassenor J, Maturana G, Baeza T, Freer J, Dural N. ZnO-catalysed photodegradation of Kraft black liquor. Journal of Photochemistry and Photobiology A: Chemistry. 1994;**78**:267-273. DOI: 10.1016/1010-6030(93)03731-U

[35] Nasr C, Vinodgopal K, Hotchandani S, Chattopadhyay AK, Kamat PV. Photocatalytic reduction of azo dyes naphthol blue black and disperse blue 79. Research on Chemical Intermediates. 1997;**23**:219-231

[36] Low GKC, Mcevoy SR, Matthews RW. Formation of ammonium and nitrate ions from photocatalytic oxidation of ring nitrogenous compounds over titanium dioxide. Chemosphere. 1989;**19**:1611-1623

[37] Nohara K, Hidaka H, Pelizzetti E, Serpone N. Processes of formation of NH_4^+ and NO_3 ions during the photocatalysed oxidation of N-containing compounds at the titania/water interface. Journal of Photochemistry and Photobiology A: Chemistry. 1997;**102**:265-272. DOI: 10.1016/S1010-6030(96)04478-4

[38] Muneer M, Bahnemann D. Semiconductor mediated photocatalysed degradation of two selected pesticide Terbacil and 2, 4, 5, Tribromoimidazole. Water Science and Technology. 2001;**4**:331-337

[39] Sokmen M, Ozkan A. Decolourising textile wastewater with modified titania: The effects of inorganic anions on the photocatalysis. Journal of Photochemistry and Photobiology A: Chemistry. 2002;**147**:77-81. DOI: 10.1016/S1010-6030(01)00627-X

[40] Nevim S, Arzu H, Gulin K, Cinar Z. Prediction of primary intermedites and the photodegradation kinetics of 3-aminophenol in aqueous TiO_2 suspensions. Journal of Photochemistry and Photobiology A: Chemistry. 2001;**139**:225-232. DOI: 10.1016/S1010-6030(01)00368-9

[41] Galindo C, Kalt A. UV–H_2O_2 oxidation of monoazo dyes in aqueous media: A kinetic study. Dyes and Pigments. 1999;**40**(1):27-35. DOI: 10.1016/S0143-7208(98)00027-8

[42] Pelizzetti E, Schiavello M. Photochemical Conversion and Storage of Solar Energy. Kluwar Acad. Publ. Dordrecht; 1991

[43] Shu HY, Huang CR. Degradation of commercial azo dyes in water

using ozonation and UV enhanced ozonation process. Chemosphere. 1995;**31**(8):3813-3825. DOI: 10.1016/0045-6535(95)00255-7

[44] Sivalingam G, Nagaveni K, Hegde MS, Madras G. Photocatalytic degradation of various dyes by combustion synthesized nano anatase TiO_2. Applied Catalysis B: Environmental. 2003;**45**:23-28. DOI: 10.1016/S0926-3373(03)00124-3

[45] Herrmann JM, Disdier J, Pichat P. Photoassisted platinum deposition on TiO_2 powder using various platinum complexes. The Journal of Physical Chemistry. 1986;**90**:6028-6034. DOI: 10.1021/j100280a114

[46] Yuexiang L, Gongxvan L, Shuben L. Photocatalytic transformation of rhodamine B and its effect on hydrogen evolution over Pt/TiO_2 in the presence of electron donors. Journal of Photochemistry and Photobiology A: Chemistry. 2002;**152**:219-228. DOI: 10.1016/S1010-6030(02)00210-1

[47] Vamathevan V, Amal R, Beydoun D, Low G, Mc Evoy S. Photocatalytic oxidation of organics in water using pure and silver-modified titanium dioxide particles. Journal of Photochemistry and Photobiology A: Chemistry. 2002;**148**:233-245. DOI: 10.1016/S1010-6030(02)00049-7

[48] Vorontsov AV, Savinov EN, Jin ZS. Influence of the form of photodeposited Pt on titania upon its photocatalytic activity in CO and acetone oxidation. Journal of Photochemistry and Photobiology A: Chemistry. 1999;**125**:113-117. DOI: 10.1016/S1010-6030(99)00073-8

Chapter 6

Selective Mono-Hydrogenation of Polyunsaturated Hydrocarbons: Traditional and Nanoscale Catalysis

Ting-An Chen and Young-Seok Shon

Abstract

Selective hydrogenation of olefins is an important process in both chemical and pharmaceutical industries. This chapter reviews intriguing catalytic studies accomplished by employing a variety of catalysts such as metal complexes, supported materials, supported metal complexes, and nanosized materials for polyene hydrogenation. In addition, new research area involving unsupported colloidal nanoparticle catalysts, which exhibit an excellent activity and selectivity toward the polyene hydrogenation is introduced. The high activity of colloidal metal nanoparticle catalysts often allows the reactions to be completed under mild conditions, at atmospheric pressure, and room temperature. These colloidal nanoparticle catalysts also offer an advantage of facile separation and multiple recycling without significant losses in activity and selectivity. This chapter provides important fundamental understandings on the influence of chemical environments (solvents, ligands, dopants, etc.) and compositions (metal complex, metals, alloys, etc.) toward the catalytic activity and selectivity of various catalysts in homogeneous, heterogeneous, and semi-heterogeneous conditions. The systematic evaluation discussed in this chapter would pave a way to further develop chemo-, regio-, and stereo-selective catalysts for polyene hydrogenation.

Keywords: selective hydrogenation, catalysis, nanocatalysis, homogeneous, heterogeneous, semi-heterogeneous

1. Introduction

Selective hydrogenation of polyunsaturated hydrocarbons including polyenes and alkynes are vitally important processes in fine chemical industries [1]. For instance, polymerization and hydroformylation reactions require a high purity of monomeric alkenes. However, light alkenes produced by catalytic cracking of petroleum often contain a high level of dienes or alkynes, making the selective hydrogenation of these compounds to monoenes in the presence of alkenes that are highly essential [2]. The removal of polyunsaturated hydrocarbons is also often important for catalysis applications, since their strong adsorption to metal surfaces would deactivate the catalyst during the reaction. In addition, the unique scent of natural polyunsaturated compounds makes them an important ingredient in perfume industry and more attentions are currently placed on the single hydrogenation products of the natural triene substrates as both perfume ingredients and pharmaceutical precursors [3, 4]. Many efforts have been carried out in the past for selective hydrogenation using either homogenous molecular catalysis or heterogeneous solid state reactions [5, 6]. With both the pros and cons of each approach, the semi-heterogeneous characteristics of soluble colloidal metal nanoparticles in addition to their large surface area to volume ratio have increased research interests on nanoparticle catalysts for selective organic reactions. This chapter reviews the up-to-date progress on the selective hydrogenation of polyunsaturated olefins using both traditional and nanoscale catalysts.

2. Traditional catalysts

2.1 Metal complexes for hydrogenation of dienes in homogeneous system

Homogeneous system using metal complex usually exhibits high reactivity for the catalytic hydrogenation of dienes [7, 8]. Schrock et al. proposed the reaction mechanism for the catalysis of homogeneous rhodium metal complex ($[RhL_n]^+$) that is efficient for the selective hydrogenation of norbornadiene (NDB). Based on the results of 1a–1d in **Table 1**, the diene reaction rate was not significantly affected by the addition of other reagents. **Table 1** also presents the effect of the size of the ligand on metal complex (R_{ene} values of 1a, 2, and 3 in **Table 1**). These results indicated that the reactivity decreases as the size of the ligand increases. This phenomenon is similar to other homogenous reactions of metal complex. Dienes could effectively chelate to the metal complex and form strong bonds with metal atoms, even under the presence of excess hydrogen gas. The strong bond formation between metal ion and diene could be visualized by the rapid color change of the complex after diene was added to the solution. The overall results indicated that the major path involves the coordination of olefins and the adsorption of hydrogen ($[RhH_2(NBD)L_n]^+$), which are followed by the addition of hydride to form Rh-alkyl intermediate ($[RhH(NBD\text{-}H)L_n]^+$). After the reductive elimination of alkyl and hydride, rhodium metal complex ($[RhL_n]^+$) is regenerated and is ready for recycle. The deuterium gas addition study for $[Rh(NBD)L_n]^+$ revealed that the substrate is chelated on metal ion by the two π bonds. Therefore, the primary hydrogenation reaction would have two deuterium on the endo side of the norbornene.

Run	Catalyst	R_{diene}	R_{ene}	Max % ene
1a	$[Rh(NBD)(PPh_3)_2]^+PF_6^-$	0.22	0.03	97
1b	1a with 3.0 mol of $HClO_4$	0.21	0.05	92
1c	1a with D_2	0.22	(b)	(b)
1d	1a with 2.0 mol of Et_3N and D_2	~0.21 (initial)[a]	(b)	80
2	$[Rh(NBD)(PPh_2Me)_2]^+PF_6^-$	0.16	0.12	97
3[b]	$[Rh(NBD)(PPhMe_2)_2]^+PF_6^-$	0.14	0.19	90

[a]*A markedly nonlinear rate was observed. The behavior was more nearly first order in olefin* ($k = 4.4 \times 10^{-4} s^{-1}$).
[b]*Catalyst precursor = 0.053 mmol.*
Data reproduced from [7].

Table 1.
The hydrogenation of norbornadiene in acetone with Rh complex (in 10.0 ml of acetone, 1.0 ml of NBD, 30.0 ± 0.5°C, 1 atm total pressure of H_2*, 0.026 mmol of catalyst precursor, R = rate in mmol/min).*

The metal complex catalyst could also be deactivated once two bonds on dienes are chelated to one metal ion. Because the metal ion with a chelated diene compound is too stable to react, the bidentate ligand as shown in **Figure 1** was essential in avoiding this deactivation by diene coordination. The catalysis results of substituted dienes by the rhodium complex with different ligands showed the influence of ligands on the catalytic selectivity for 1,2- and 1,4-addition products. The comparisons of results indicated that diphos and arphos favor the 1,4-addition product (80–90%), while dpea favors the 1,2-addition product (~80%). Between diphos and arphos, arphos exhibits slightly higher selectivity toward the 1,4-addition product. The catalytic reaction of 1,4-cyclohexadiene begins with the isomerization to 1,3-cyclohexadiene. The produced 1,3-cyclohexadiene, however, would not dissociate from the metal ion, forming $[Rh(diene)L_n]^+$, due to the strong bond formation between metal ion and diene compound. The monoene compound would dissociate from metal ion after its formation, because it forms a weakened bond with metal ion. The high conversion yields (>98%) of these catalytic reactions indicated the high reactivity of the $[Rh(diene)L_n]^+$ catalyst in general.

Frankel et al. showed that other metal complex catalysts such as chromium complex, methyl benzoate-$Cr(CO)_3$, also favors the 1,4-addition reduction of dienes for the hydrogenation of 1,3- and 2,4-hexadiene (70–90%) [8]. In contrast, the formation of 1,4-addition hexene product was accompanied with the major formation of conjugated 1,3- and 2,4-diene products for the catalytic hydrogenation of 1,4- and 1,5-hexadiene indicating that the reaction would most likely involves the first isomerization of isolated dienes to conjugated intermediates. The catalytic reactions of conjugated dienes with methyl substituted group(s) also mostly produced the 1,4-addition hydrogenation products as shown in **Table 2**. The position of substituted methyl group would not have a major effect on the catalytic activity except the case for 2,5-dimethyl-2,4-hexadiene, which exhibits low reactivity due to the large steric interference of four methyl groups. The difficulty in generating the product with cis-conformation, which chromium complex catalyst favors, is considered to be the main reason. The 1,2-addition hydrogenation product generated from the catalytic reaction of 4-methyl-1,3-pentadiene is also turned out to be the 1,4-addition product involving H shift. The isomerization of 1,4-cyclohexadiene was also more favorable than the 1,4-addition hydrogenation, forming 1,3-cyclohexadiene as the major product. This result also proves that the isomerization would take place prior to the hydrogenation. When the reaction temperature is increased to 170°C, the hydrogenation product becomes the major (thermodynamic) product. The high stability of 1,3-cyclooctadiene also reduces the reactivity

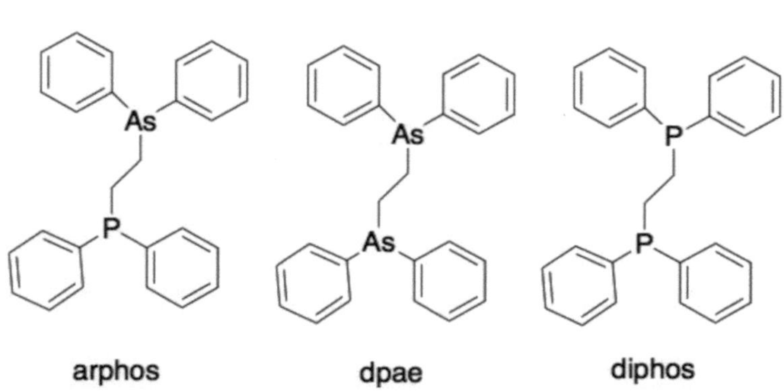

Figure 1.
Different bidentate ligands for diene catalysis reaction by rhodium complex [7].

of substrate at the lower temperature resulting in low yield for the hydrogenation product.

Regioselective asymmetric monohydrogenation of 1,4-dienes has been studied using various organometallic catalysts including ruthenium, rhodium, and iridium complexes with N or P binding chiral ligands [9–12]. The iridium catalysts exhibited excellent enantioselectivity for the hydrogenation of disubstituted cyclohexadienes as shown in **Figure 2** below [12]. The catalytic reactions produced mono-hydrogenation products shown below as major products in the yields ranging from 45 to 99% depending on the structure of O-R group. Tetrahydropyranyl acetal (THP) and triisopropyl silyl ether (TIP) resulted in 99% monohydrogenation selectivity. The enantioselectivity of these two groups were 83 and 97%ee, respectively, indicating highly efficient regio- and enantioselectivity of Ir catalyst for the synthesis of silyl protected enol ethers. Oxidation of these chiral enol ethers led to the corresponding chiral α,β-unsaturated ketones.

Dienes, 9.5 mmol	k, h^{-1}	Products (% at 6 h)
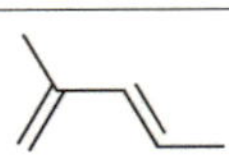	0.69	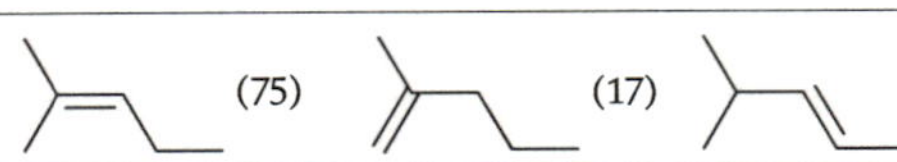 (75) (17) (8)
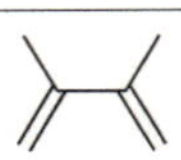	0.64	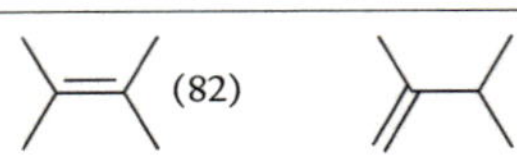 (82) (18)
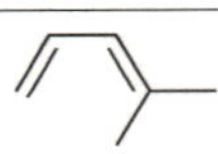	1.1	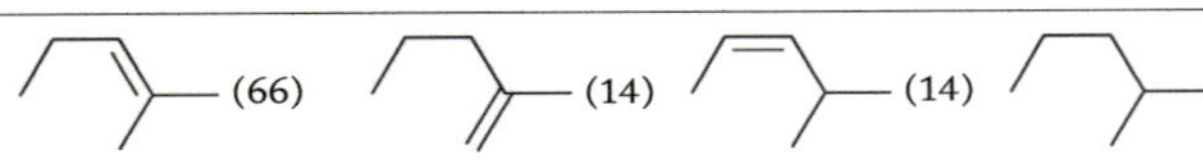 (66) (14) (14) (6)
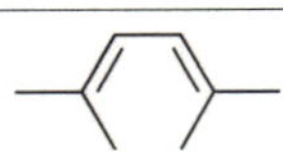	0.007	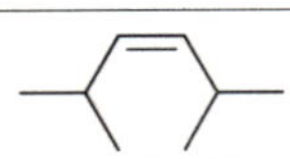 (5)
	0.1^b	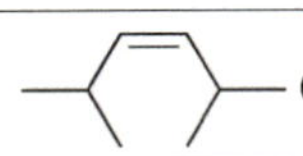 (45)

b At 175°C.
Data reproduced from [8].

Table 2.
Catalytic hydrogenation of methyl-substituted dienes with 0.5 mmol chromium complex (solvent: n-pentane, 50 ml; temperature: 160 °C; initial H_2 pressure: 30 atm.).

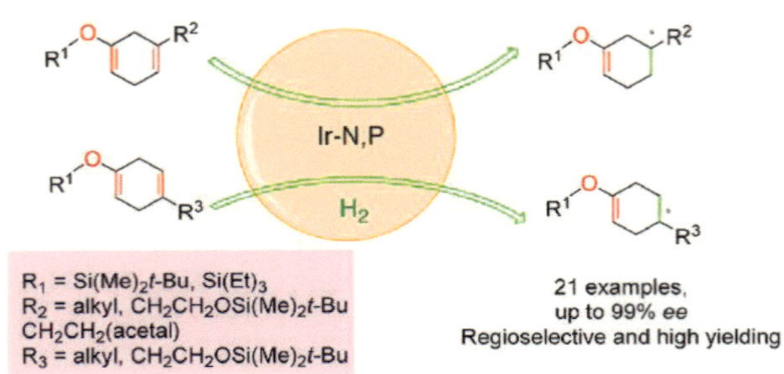

Figure 2.
Asymmetric hydrogenation using iridium metal complex (0.5 mol% Ir catalyst, $PhCF_3$, K_3PO_4, H_2, rt., 12 h). Reproduced from [12] with permission from the American Chemical Society.

2.2 Supported materials for the hydrogenation of dienes in heterogeneous system

In terms of reactivity, heterogeneous catalysts are usually less reactive than homogeneous systems. However, heterogeneous catalysts only require simple separation processes for purification and can be more easily recycled compared to homogeneous catalysts. Therefore, many research groups have been working on advancing fundamental understanding on the structure/property relationship and technological applications of heterogeneous catalysts in the past decade. Many of these supported metal catalysts are in fact in nanoscale dimensions, but many earlier catalysis studies did not attempt detailed characterizations on material sizes and their distributions on the supports. Both materials with and without well-defined sizes and structures including the catalysts reported as nanoparticulate materials are discussed here as traditional catalysts.

2.2.1 Supported metal catalysts: different strategies to modify activity

Selective hydrogenation of 1,3-butadiene was studied using graphite-supported palladium and platinum and the influence of FeCe alloying to these heterogeneous catalysts was investigated [13]. The results showed that the mono-hydrogenation and subsequent isomerization to 2-butene takes better place when the alloying was limited to less than 1/20 (**Figure 3**). The monohydrogenation selectivity was ascribed to the depletion of hydrogen atoms away from palladium surfaces by spill over to alloyed metal surfaces. The overall catalytic activity has also been increased by alloying of FeCe to Pd or Pt catalysts, indicating the activation of FeCe by spill over hydrogen.

Similarly, alumina-supported palladium catalysts doped with either tin or silver were tested for the selective hydrogenation of 1,5-hexadiene and 1,3-hexadiene [14]. Palladium on alumina itself produced mono-hydrogenation products from both 1,5-hexadiene and 1,3-hexadiene with a high selectivity even at full conversions. However, the selectivity for 1-hexene (or 3-hexene from 1,3-hexadiene) over the isomerized 2-hexene (trans > cis) from 1,5-hexadiene started to decrease at conversions higher than 80%. The addition of tin or silver tends to significantly increase the selectivity for 1-hexene, but with the loss of overall activity for mono-hydrogenation. This indicated that the addition of doping metal causes a geometric dilution of active Pd adsorption sites for both double-bond isomerization and hydrogenation.

Sulfidation of supported Pd catalysts has also been identified as an efficient way to increase the selectivity for mono-hydrogenation of dienes [15]. Supported palladium sulfide catalysts could be prepared by the addition of H_2S or Na_2S or the treatment with fuming sulfuric acid [16]. The produced palladium sulfide (Pd_4S) catalysts deposited on carbon nanofibers exhibited the mono-hydrogenation activity in the gas-phase butadiene reduction producing butenes of various forms in good yields (99% of butenes at 100% conversion: the selectivities among various butenes are not reported). In contrast to Pd metal-based catalysts, this Pd_4S catalyst presented high stability under reaction conditions while having significant activity and appropriate selectivity for partial hydrogenation of dienes.

Thiolate self-assembled monolayers deposited on Pd/Al_2O_3 catalysts could also direct the catalytic activity of heterogeneous systems for fatty acid diene hydrogenation as shown in **Figure 4** [17]. In comparison, the uncoated Pd/Al_2O_3 catalyst produced the fully hydrogenated fatty acids under the same hydrogenation condition. This selectivity is attributed to steric effects between thiolate monolayers and

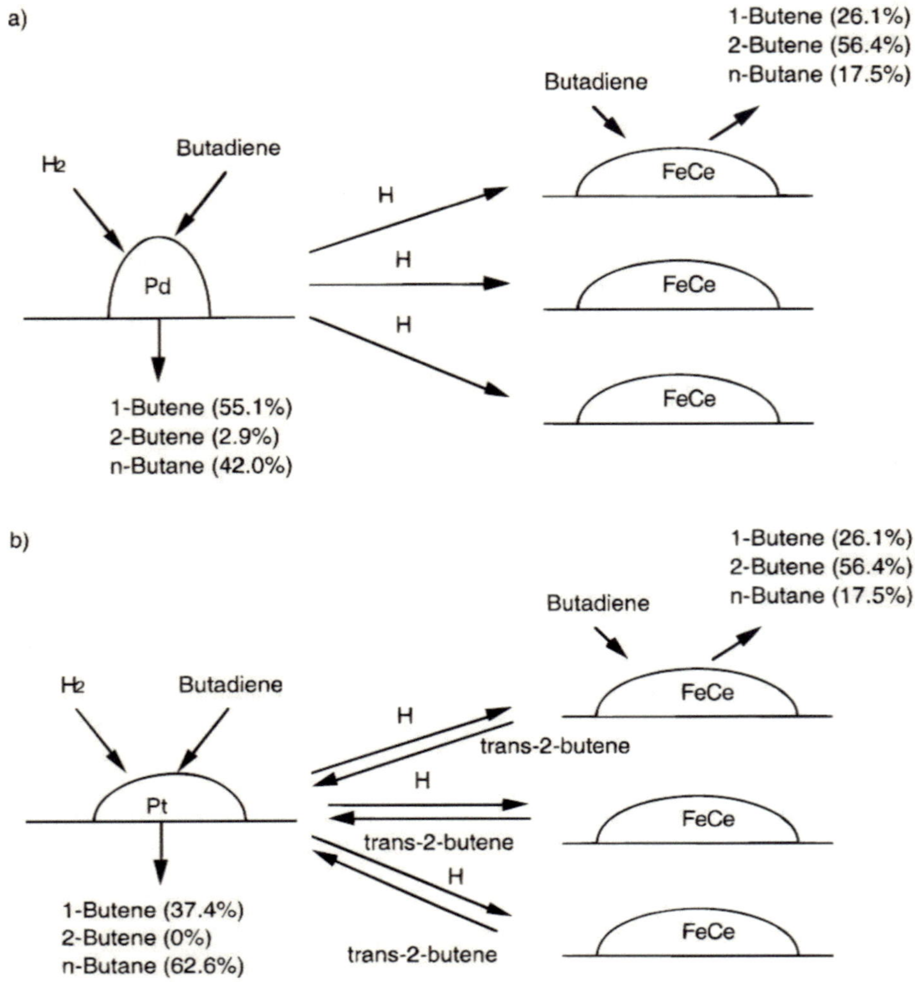

Figure 3.
The working hypothesis of physical mixtures for 1,3-butadiene hydroisomerization. (a) Pd containing mixtures and (b) Pt containing mixtures. Reproduced from [13] with permission from the American Chemical Society.

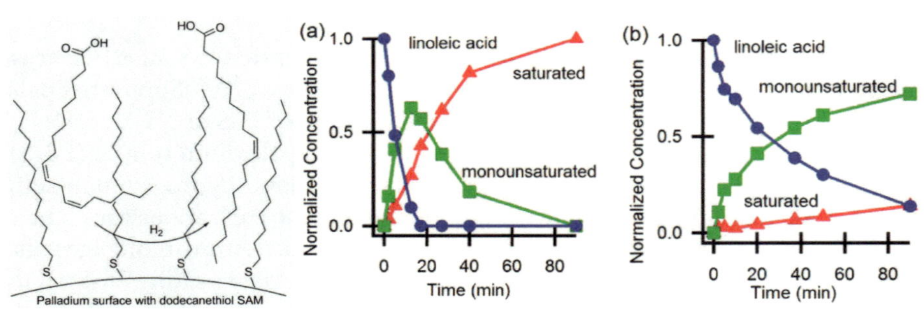

Figure 4.
Kinetic data for linoleic acid hydrogenation over Pd/Al_2O_3 at 30°C and 6 bar H_2. (a) Uncoated Pd/Al_2O_3 and (b) dodecanethiol-coated Pd/Al_2O_3. Reproduced from [17] with permission from the American Chemical Society.

fatty acid reactants based on the kinetic studies and ligand chain length studies reported in this work. The influence of ligand chemical functionality was also investigated in this study. The results showed that unlike hydrophobic alkanethiolate ligand modifiers, the hydrophilic thioglycerol modifiers strongly inhibited the catalytic activity of Pd/Al_2O_3 surfaces.

2.2.2 Supported metal nanoparticle catalysts: the nano effects

More and more researchers consider nanoparticles as a better option for catalytic reactions due to their high surface area per volume characteristics. In the area of heterogeneous catalysis, the complete analyses of catalyst sizes, compositions, and distributions are now required and many well-known active solid-state catalysts including carbon-supported Pd are found to be actually in nanoscale. With the advancement of nanomaterials synthesis and characterization, the nanoscale catalysts are now designed and prepared to tune their activities for desired applications including diene hydrogenation. For example, Pd nanoparticles stabilized with dendrimers (polypropylenimine, PPI) deposited on a silica surface are used for catalysis application (**Figure 5**) [18]. The internal amine functional groups on PPI dendrimers are used as a ligand to encapsulate Pd nanoparticles and the external amine groups help grafting the dendrimers on a polyamine-modified silica surface to form the immobilized dendrimer catalyst composite. The dendrimers around the under-deposited nanoparticle increase the selectivity of the Pd nanoparticles and decrease the Pd metal leaching. The immobilized dendrimer catalyst reveals higher reactivity for the selective hydrogenation of dienes than the traditional heterogeneous catalysts. In this study, Karakhanov et al. further discussed the effects of size and substituent pattern of the substrate, 2,5-dimethyl-2,4-hexadiene, during the catalytic hydrogenation (**Table 3**). Since both C=C double bonds in 2,5-dimethyl-2, 4-hexadiene are internal and highly substituted at C2 and C5 positions, the rate of the reaction is relatively slow but the overall reactions result in the high yield of thermodynamic 1,4-addition product.

Instead of using modifier or poisoning agents to change the catalytic activity of heterogeneous metal substrates, the modification of support materials to induce the steric-related selectivity has been successfully attempted [19]. By overcoating Pd nanoparticle catalyst with porous alumina using atomic layer deposition, Yi et al. could produce a highly stable (against deactivation) and selectivity for mono-hydrogenation of 1,3-butadiene to butenes (**Figure 6**). The selective hydrogenation worked well even in the presence of excess propene. The alumina overcoat clearly suppressed the conversion of prepene to propane very efficiently while maintaining ~100% butenes selectivity with ~100% 1,3-butadiene conversion. This is attributed

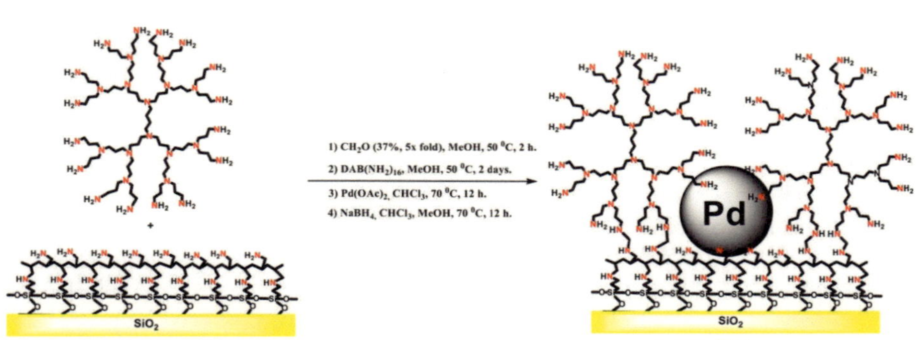

Figure 5.
Polypropylenimine (PPI)-modified palladium nanoparticle catalyst composite. Reproduced from [18] with permission from the American Chemical Society.

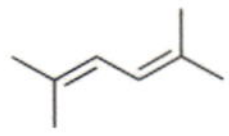 Entry	P, MPa	t, h	Sub/Pd	Conv.,%	Product distribution 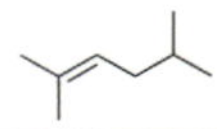	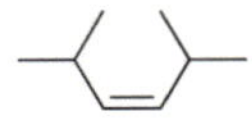	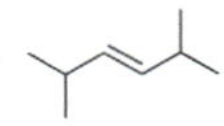	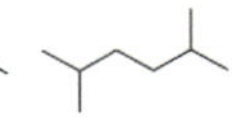
1	1	1	3680	100	16.5	22	58	4
2	3	1	3680	100	17.5	23.5	53	5
3	1	0.25	3680	100	20	26	51	4.5
4	1	0.25	7360	91.5	9	64.5	25.5	1

Data reproduced from [18].

Table 3.
The catalytic reactions of 2,5-dimethyl-2,4-hexadiene with Pd nanoparticle composite in 2 mL toluene at 70°C and under 3 MPa H_2.

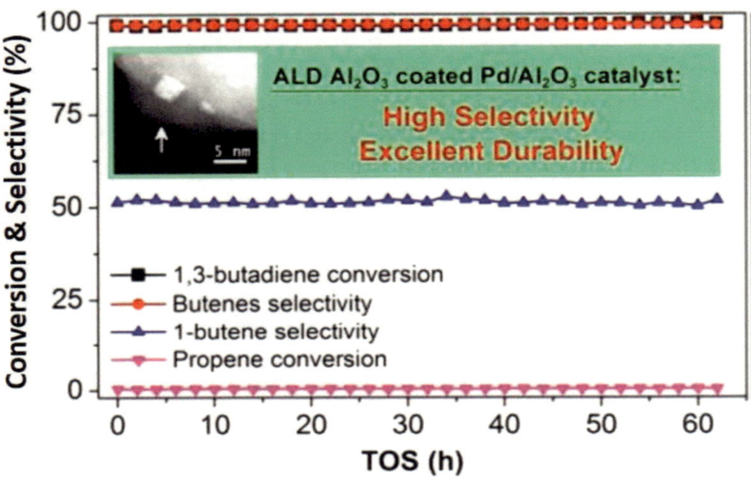

Figure 6.
Catalytic property of microporous alumina-coated Pd/Al_2O_3 using atomic layer deposition (ALD) for 1,3-butadiene hydrogenation in the presence of an excess propene. Reproduced from [19] with permission from the American Chemical Society.

to the confinement effect within the micropores of the alumina overcoat and the stronger adsorption of 1,3-butadiene than alkenes on Pd nanoparticle catalysts.

Mesoporous carbon films as support to control the activity of Pd catalyst also reported for the selective hydrogenation of 1,3-butadiene [20]. The material synthesis involves the co-deposition of small polymeric carbon clusters, structure filling agents, and Pd ions on a substrate (**Scheme 1**). Thermal treatments converted these hybrids into graphitized microporous carbon with active Pd catalysts. These catalysts were highly active in the gas-phase mono-hydrogenation of 1,3-butadiene to butenes. The major product for these catalytic systems is 1-butene at ~50% selectivity among butene isomers, which is very similar to the catalytic selectivity of porous alumina-coated Pd catalysts described above.

Bimetallic Au-Pd alloy catalysts with low amount of Pd were prepared by either co-deposition–precipitation or co-impregnation procedure [21]. This approach is especially beneficial considering the low usage of somewhat toxic Pd metals. These bimetallic catalysts could selectively hydrogenate 1,3-butadiene in the presence of propene. By changing the Au/Pd ratio, the catalytic activity of bimetallic catalysts could be further controlled. The overall selectivity among butene isomers also depended on the reaction temperatures. At the lower temperature, 1-butene was

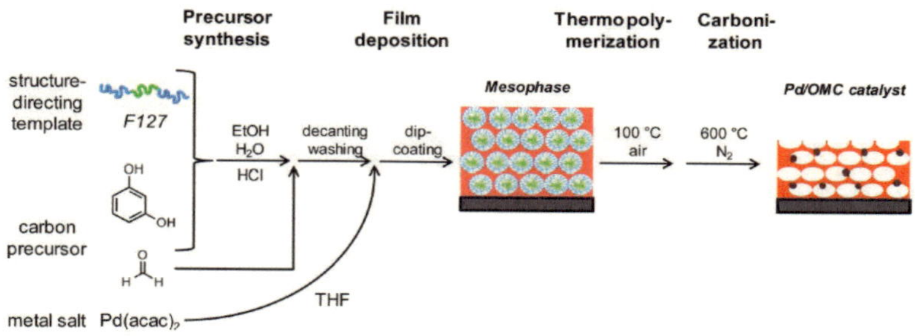

Scheme 1.
Illustration of the synthetic approach to mesoporous Pd/OMC catalyst films. Reproduced from [20] with permission from the American Chemical Society.

obtained as major product, reaching up to ~76% selectivity. At the higher temperature, the selectivity for thermodynamically stable trans-2-butene increased with the best selectivity at ~60%.

2.2.3 Supported metal complexes: bridging homogeneous catalysis with heterogeneous catalysis

To increase the site efficiency of heterogeneous catalysis, metal complexes with 100% site efficiency are deposited on to the support surface. For example, Zweni et al. integrated palladium metal ion complex on silica gel supported dendron ligands with different generations (**Figure** 7) [22]. The catalytic reactions of 1,3-cyclohexadiene in various solvent systems are investigated to see the influence of solvent characteristics on the reactivity and selectivity of the catalyst. Methanol is found to be the optimized solvent system for this catalyst based on the selectivity to cyclohexene by mono-hydrogenation.

This silica-supported PAMAM-palladium complex would exhibit different catalytic reactivities and selectivities for hydrogenation of dienes depending on the dendrimer sizes and linker chain lengths (**Table 4**). Overall, the similar selectivity is observed for Entries 1 (shorter reaction time) and 2 (longer reaction time). Generation 0/complex 1 and generation 2/complex 6 exhibit higher reactivity than other generation/linker combinations. Entries 2, 4, and 6 show the results of the catalytic reactions by G1 catalysts at the first 30 min, which suggest the high initial selectivity of these catalysts toward cyclohexene. With the increased reaction time,

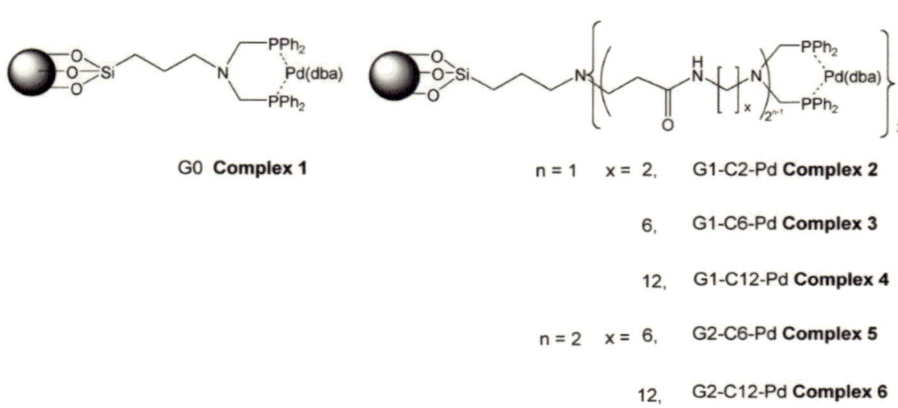

Figure 7.
Silica-supported PAMAM-palladium complex catalyst. Reproduced from [22] with permission from WILEY.

Entry	Catalyst	Time (h)	Conv. (%)	 (%)	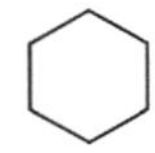 (%)	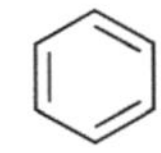 (%)
1	G0, **C1**	1.75	>99	76	24	—
2	G1-C2, **C2**	0.5	21	>99	—	—
3		20	>99	88	10	2
4	G1-C6, **C3**	0.5	11	>99	—	0
5		20	72	97	1.5	1.5
6	G1-C12, **C4**	0.5	15	>99	—	—
7		20	20	>99	—	—
8	G2-C6, **C5**	0.5	N.R	—	—	—
9		20	>99	73	11	6
10	G2-C12, **C6**	0.5	24	>99	—	—
11		5	>99	80	14	6

GC and NMR are used to monitor the reaction.
Data reproduced from [22].

Table 4.
Hydrogenation of 1,3-cyclohexadiene with silica supported PAMAM-palladium complex catalyst (5.25mmol 1,3-hexadiene, 10 μmol Pd catalyst complex, 5 mL methanol, pressurized glass autoclave to 14 psi H_2, 25°C).

the catalysis results shown in Entries 3, 5, and 7 indicate the increased conversion of reactants with the formation of some minor full hydrogenation products. The results suggest that the primary product, cyclohexene, would compete with the diene reactant for the catalytic active sites. When the concentration of 1,3-cyclohexadiene is high at the beginning of the reaction, diene would dominate the adsorption on the active sites and the reaction would maintain a good selectivity to cyclohexene. With the increased concentration of cyclohexene at the later stage of the reaction, the adsorption on active catalytic sites starts to take place. The reactions are generally slower for G1 catalysts compared to G0 catalyst especially with longer linkers. The G2-C12 catalyst, however, exhibits good activity and selectivity toward cyclohexene indicating the importance of right combination between dendrimer generations and linker lengths. A trace amount of benzene was also observed for some of the catalysts. The catalytic reactions of other acyclic dienes also are investigated to understand the effects of substrate structure. In general, the 1,2-addition hydrogenation of less hindered C=C is the most favorable compared to that of more hindered C=C, 1,4-addition hydrogenation, and full hydrogenation.

Zeolite- and magnesium oxide-supported molecular rhodium complexes are also tested for the hydrogenation of 1,3-butadiene [23]. The selectivity for mono-hydrogenation increases when the Rh species nucleation decreased from several atoms to dimeric clusters. The poisoning with CO ligands further increases the mono-hydrogenation selectivity, especially when electron donating MgO is used as a support (>99% selectivity for mono-hydrogenation as shown in **Figure 8**). This is attributed to limiting the activity for H_2 dissociation and preventing the additional hydrogenation to butane.

2.3 Metal complexes and supported materials for the hydrogenation of trienes

Selective hydrogenation of triene is also an important topic for fine chemicals and pharmaceutical industries [3]. Myrcene with one isolated C=C bond and two

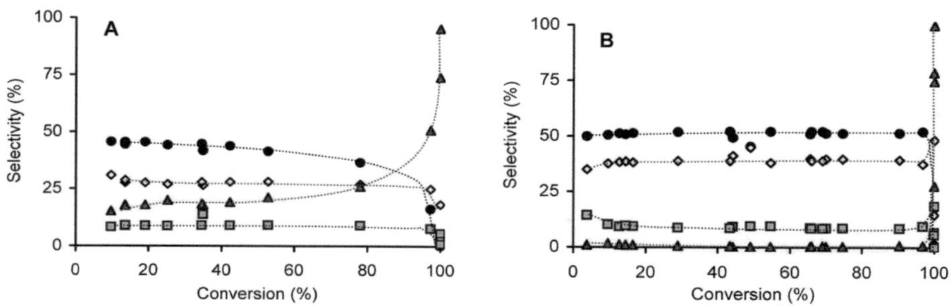

Figure 8.
Selectivity plots for MgO-supported rhodium dimers in the absence (A) and in the presence (B) of CO ligands in the hydrogenation of 1,3-butadiene (filled circle: 1-butene, open diamond: trans-2-butene, blue square: cis-2-butene, red triangle: butane). (Reactions condition: 2 vol % 1,3-butadiene, balanced with H_2, total pressure = 1 bar, room temp) Reproduced from [23] with permission from the American Chemical Society.

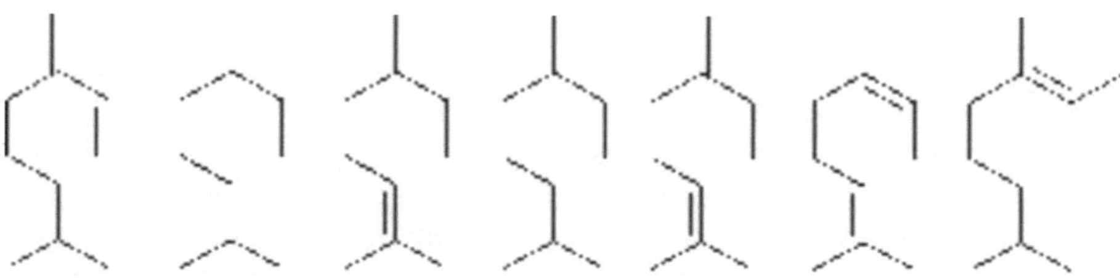

Figure 9.
Reactant (1) and potential products (2–7) for myrcene hydrogenation. Products 4–7 are dienes. Reproduced from [24] with permission from the Sciencedirect.

conjugated C=C bonds can be synthesized by the pyrolysis of β-pinene in nature. Despite the availability of myrcene, only little success in selective hydrogenation of the substrate to diene have been made. Gusevskaya et al. reported the hydrogenation of myrcene either by using metal complex ion or heterogeneous sol-gel catalysts [24, 25]. For 10% palladium on carbon (Pd/C) under 20 atm of H_2 and cyclohexane solvent system at 80°C, full hydrogenation reaction would take place within 30 min, and the catalytic system does not have any selectivity toward diene or monoene. On the other hand, the transition metal complexes of $[RuCl_2(CO)_2(PPh_3)_2]$, $[RhH(CO)(PPh_3)_3]$, $[IrCl(CO)(PPh_3)_2]$, and $[Cr(CO)_6]$ show the capability to form diene products in relatively good yields (**Figure 9** and **Table 5**) [24].

From Entries 1–4 in **Table 5**, the reactivity of the metal complexes under the same condition turns out to be Ru < Cr < Ir < Rh. Ru and Rh complexes show slightly higher selectivity toward monohydrogenated products, dienes, than chromium and iridium complexes. Rh complex is further tested by adding extra PPh_3 ligand in the reaction (Entry 5). The presence of extra PPh_3 slows the reaction down requiring higher reaction temperature, but increases selectivity toward dienes. The similar selectivity of myrcene hydrogenation is achieved by simply changing solvent from cyclohexane to benzene even at the lower reaction temperature of 80°C. The reaction only produces a trace amount of double or full hydrogenation products. However, the selectivity among different dienes (Entries 4–7) is still poor.

The first hydrogenation of myrcene, a triene, takes place at the conjugated diene group instead of the isolated and hindered alkene group. The 1,2-addition of conjugated diene produces either compound 4 or compound 5. The 1,4-addition of conjugated diene involving Pd-allyl intermediates produce compounds 6 and 7, the

Run	Catalyst	Time (min)[a]	T (°C)	S (%)[b]	Product distribution (%)					
					2	3	4	5	6	7
1	$[RuCl_2(CO)_2(PPh_3)_2]$	110	100	83	1	16	7	32	9	35
2	$[Cr(CO)_6]$	45	100	74	4	22	8	26	29	11
3	$[IrCl(CO)(PPh_3)_2]$	15	100	76	4	20	8	22	33	13
4	$[RhH(CO)(PPh_3)_3]$	5	100	87	4	9	13	24	34	16
5	$[RhH(CO)(PPh_3)_3]^c$	24	140	96	tr.	4	14	31	26	25
6	$[RhH(CO)(PPh_3)_3]^d$	60	80	98	tr.	2	15	27	34	22

[a]Reaction time necessary for ca. 80% conversion.
[b]Selectivity for monohydrogenated products 4–7 at ca. 80% conversion.
[c]PPh_3 was added (0.17 mmol).
[d]Benzene was used as a solvent.
Data reproduced from Ref [24].

Table 5.
The catalytic reaction of metal ion catalyst with myrcene, compound 1, under 20 atm H_2 and cyclohexane solvent.

E-Z isomers. Based on the results in **Table 5**, the major diene products are compound 5 and 6. This is due to the higher reactivity of terminal alkene in myrcene, which undergoes the coordination of the primary alkene group followed by the hydrogen addition. These isolated diene products 4–7 are further hydrogenated to monoene 3 with the addition of hydrogen to less sterically hindered alkene. When the reaction is continued, the yield for full hydrogenation product 2 constantly increases. Since there are several pathways for the hydrogenation of myrcene and the reactions generate many different mono- and di-hydrogenation products, this reaction is extremely difficult to control and hard to predict with regarding the overall selectivity. However, there are some correlations between the yields of products and the kinetics/thermodynamics of intermediates and products. For example, the kinetic reactivity among diene products should follow the ensuing order: 4 > 5 > 6 > 7.

For the hydrogenation of myrcene by sol–gel Pd/SiO_2 catalyst, the selectivity for dienes is higher than that of the metal complex catalyst [25]. Robles-Dutenhefner et al. used three different temperatures for the reactions, the catalysis results show

Run	Catalyst (wt. %)	T (°C)	Time (min)	Conv. (%)	S (%)[a]	Product distribution (%)				
						2	3	4	5	6 + 7
1	1% Pd/SiO_2/300°C (0.5)	80	15	100	0	100				
4	1% Pd/SiO_2/1100°C (0.5)	80	15	75	100			20	18	62
			60	96	94	1	5	16	16	62
5	1% Pd/SiO_2/1100°C (0.5)	100	15	80	99		1	18	15	66
6	1% Pd/SiO_2/1100°C (2.5)	80	15	75	98	1	1	18	15	65
8	3% Pd/SiO_2/1100°C (0.5)	80	60	96	97	1	2	20	15	62

[a]Selectivity for monohydrogenated products 4–7.
Data reproduced from Ref [25].

Table 6.
Hydrogenation of myrcene catalyzed by Pd/SiO_2 in cyclohexane under 20 atm H_2.

that the reaction temperature would greatly affect the reactivity of the Pd/SiO_2 catalysts. BET surface area analysis shows the change in synthesis temperature that would cause some variations in the pore size of the catalyst. Since the pore is created by the presence of organic solvent during the synthesis process, the high temperature at or above 300°C would generate the largest pore size due to increased gas pressure. However, unstable gaseous environment at higher temperature of 1100°C causes the collapse in its pore resulting in the decreased surface area and pore size.

The catalysis results for myrcene hydrogenation in **Table 6** show that the selectivity toward dienes by mono-hydrogenation is higher than 90%. The Pd/SiO_2 catalyst with a larger pore created at 300°C converts the reactant to saturated organic compound without forming any alkene. Due to the higher surface energy of this catalyst with a large pore, the substrates are readily adsorbed on the catalyst surface and undergo full hydrogenation. For the Pd/SiO_2 heated at 1100°C, the lower surface energy of the catalyst makes the activity to be decreased. The hydrogenation of isolated alkene would become unfavorable and the high selectivity to form dienes from myrcene is observed for this Pd/SiO_2 heated at 1100°C. The ratio of the E/Z isomer (compound 6 and 7) is 0.176.

3. Semi-heterogeneous nanoscale catalysts

3.1 Semi-heterogeneous colloidal nanoparticles in ionic liquids

Colloidal nanoparticle catalysts are considered semi-heterogeneous due to their homogeneous characteristics (kinetic efficiency) accompanied by their heterogeneous surface property. Semi-heterogeneous catalyst can be benefited from the advantages that both homogenous and heterogeneous catalysts have. Nanoparticle catalysts can have a higher catalytic activity than their bulk and heterogeneous counterparts, especially with nano-effects of high surface area to volume ratio. They can also be easily separated from the products and recycled similar to other heterogeneous catalytic systems.

Semi-heterogeneous nanoparticle catalysts used for the hydrogenation of dienes show the relatively good selectivity and reactivity even compared to the traditional homogeneous catalysts. Ionic liquid-stabilized Pd nanoparticle catalyst shown in **Figure 10** is one of those examples used for diene hydrogenation [26]. The thin film of ionic liquid with the hydrophobic anion PF_6^- on the Pd nanoparticle surface (Pd/**sgPF$_6$**) enhances the solubility in methylene chloride and the selectivity of the catalyst. The ionic liquid on the surface acts like a cage to control which substrate would pass through the liquid ion film and reach the particle surface for catalytic reaction. The liquid ion film on catalyst creates the non-equilibrium environment, so that the rate determining step could easily be observed.

Several conjugated and isolated diene compounds are tested for the catalytic hydrogenation using Pd/**sgPF$_6$**, the hydrophobic ionic liquid nanoparticle catalyst (**Table 7**). The results show that the conjugated cyclodienes (Entries 1 and 3) are much more reactive than the isolated cyclodienes (Entries 2 and 4). The catalytic reactions of cyclohexa-1,3-diene **8** and cycloocta-1,3-diene **13** produce mono-hydrogenation products almost quantitatively. Once the monoene is formed, the further hydrogenation does not take place due to the weaker adsorption of monoenes on Pd nanoparticle surface. The adsorption of dienes on Pd nanoparticle surface by two π bonds coordination is much more preferred because Pd atom is electron deficient. In addition, the mono-hydrogenation products have lower

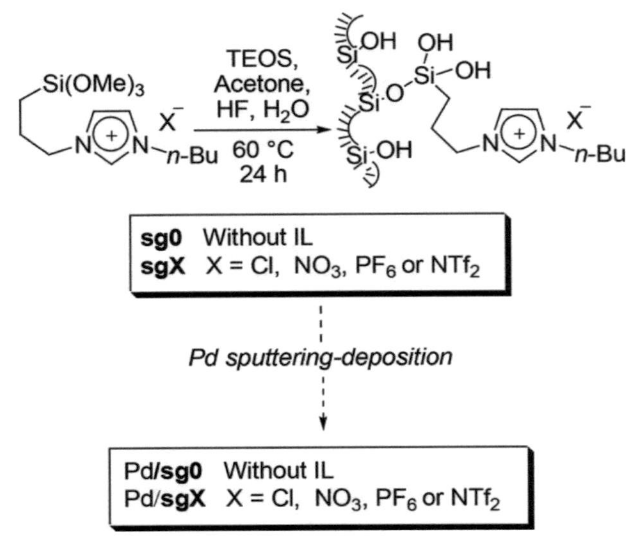

Figure 10.
Synthesis of ionic liquid hybride palladium nanoparticle by sputtering-deposition. Reproduced from [26] with permission from the American Chemical Society.

solubility in ionic liquid than diene reactants, which readily transfer through the liquid ion film. This makes monoene to be evicted out from the ionic liquid inhibiting the second hydrogenation. For isolated cyclohexa-1,4-diene **12** and cycloocta-1,5-diene **16**, the reactivity and selectivity to generate monoene are similar, but the overall conversion yields are extremely low. The low reactivity of monoene compounds is confirmed from the catalytic reactions of cyclohexene (Entry 5) and cyclooctene (Entry 6). Both substrates are unreactive for hydrogenation condition. The reaction of unsymmetrical cyclohexa-1,3-diene **17** (Entry 7) results in high conversion yields for mono-hydrogenation products with a preference for the hydrogenation of less hindered alkene group. Due to the steric hindrance of methyl and isopropyl groups, the 1,4-hydrogenation product is not formed. For the lager-conjugated dienes, the catalytic reactivity decreases with increased steric hindrance of alkyl substituent groups (Entries 9–11). However, the opposite trend for selectivity toward internal alkene is observed. As the number of alkyl substituents around C=C bonds increases, the yield for thermodynamic 1,4-hydrogenation product also increases. Moreover, the deuterium gas studies also show the actual mechanism for the cyclohexa-1,3-diene hydrogenation by ionic liquid catalyst. The conversion from cyclohexa-1,3-diene to cyclohexene is mainly through the 1,2-hydrogenation reaction with 92% selectivity and the ratio of 1,2-addition/1,4-addition around 1.7.

3.2 Semi-heterogeneous dendrimer-encapsulated metal nanoparticles

The catalytic property of polypropylenimine (PPI)-Pd nanoparticle hybrids is examined by the hydrogenation of isoprene substrate (**Table 8**) [27]. The reaction generates 2-methyl-2-butene (1,4-addition product) as the major product and 3-methyl-1-butene and 2-methyl-1-butene (1,2-addition products) as the minor products. The selectivity of the catalytic reaction is dependent upon the pressure of applied hydrogen gas and the ratio of substrate and catalyst. Higher hydrogen pressure and low substrate/catalyst ratio would result in decreased selectivity for monohydrogenation product. PPI dendrimer would enhance the catalytic selectivity to form monoene because it would limit the adsorption of the primary monoene

Entry	Diene	TOF (Conv.)[a]	Products (Selectivity/%)		
1	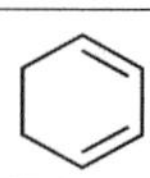 **8**	3.0 (>99)	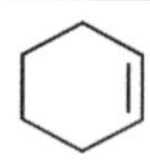 **9** (97)	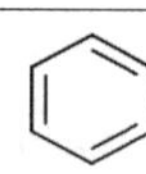 **10** (2)	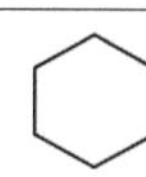 **11** (1)
2	 **12**	0.1 (<2)	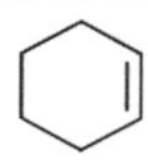 **9** (>99)	 **11** (<1)	
3	 **13**	13.0 (>99)	 **14** (>99)	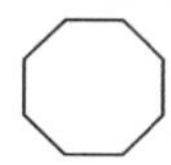 **15** (<1)	
4	 **16**	0.3 (<4)	 **14** (>99)	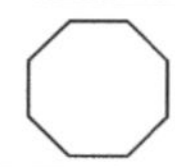 **15** (<1)	
5	 **9**	–(<1)	—		
6	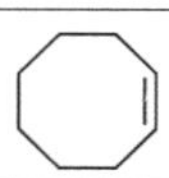 **14**	–(<1)	—		
7	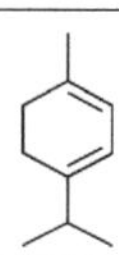 **17**	6.2 (>99)	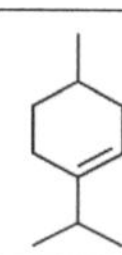 **18** (67)	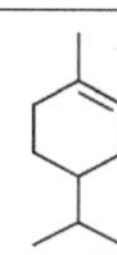 **19** (33)	
9	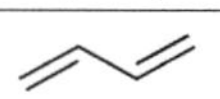 **20**	11.6 (>99)	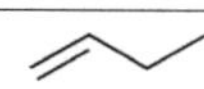 **21** (36)	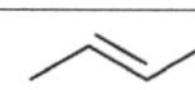 **22** (54)	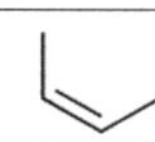 **23** (10)
10	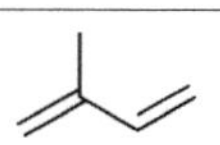 **24**	3.7 (>99)	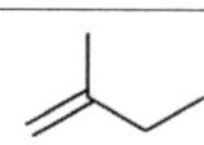 **25**(18)	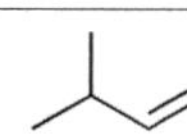 **26** (7)	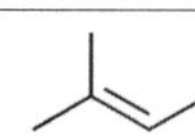 **27** (75)
11	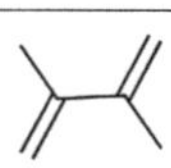 **28**	0.7 (>99)	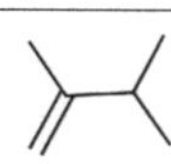 **29** (19)	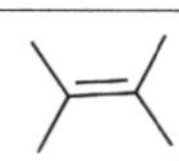 **30** (81)	

Conversion determined by GC.
[a]TOF = mol substrate converted/(mol Pd surface • s).
Data reproduced from [26].

Table 7.
Selective hydrogenation of dienes by Pd/sgPF$_6$ catalyst under optimized conditions (Reaction condition: Pd/ ***sgPF$_6$*** *catalyst (0.1 μmol Pd), substrate/Pd = 5000, 10 mL of CH_2Cl_2, 4 bar H_2, 40 °C and 250 rpm).*

product on the Pd surface and minimize the further hydrogenation to 2-methylbutane.

Other possible reasons for the high selectivity toward 1,4-addition product are also proposed. First, the initially produced 1,2-addition products could be isomerized to 1,4-addition product due to the higher stability of 2-methyl-2-butene. Second, the methyl group on isoprene would affect the stability of π-allyl intermediate (**Figure 11**) [6]. The intermediate A would generate two different products, 3-methyl-1-butene or 2-methyl-2-butene. In regards to the steric effect of intermediate A at C2 and C4, C4 has less substituted groups around C=C compared to C2. Therefore, it is easier for the second hydrogen to add on C4, and increase the yield of 2-methyl-2-butene. For the intermediate B, it is also easier for hydrogen atom to transfer to C1 since C3 is relatively more hindered than C1, which results mostly in 2-methyl-2-butene. The steric hindrance also directly influences the relative yield of 3-methyl-1-butene (higher) compared to that of 2-methyl-1-butene (lower).

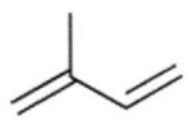 Entry	P, MPa	t, h	Sub/ Pd[a]	Conv., %	Selec. on alkene, %	Product distribution 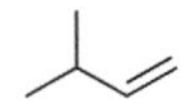	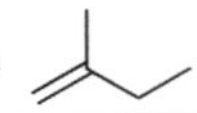	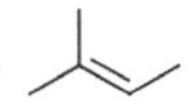	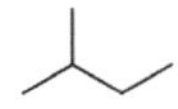
1	1	1	3680	100	96	16.5	22	58	4
2	3	1	3680	100	95	17.5	23.5	53	5
3	1	0.25	3680	100	95.5	20	26	51	4.5
4	1	0.25	7360	91.5	99	9	64.5	25.5	1
5	3	0.25	7360	100	91	16.5	24	50.5	9
6	3	0.25	14,720	100	93.5	20.5	25	48	6.5
7	3	0.25	22,080	100	94	22	24.5	47.5	6
8	3	0.25	66,240	83	98.5	19	30	49	2

[a]Mol/mol ratio.
Data reproduced from [27].

Table 8.
The catalytic reactions of isoprene with Pd nanoparticle composite in 2 mL toluene at 70°C and under 3 MPa H_2.

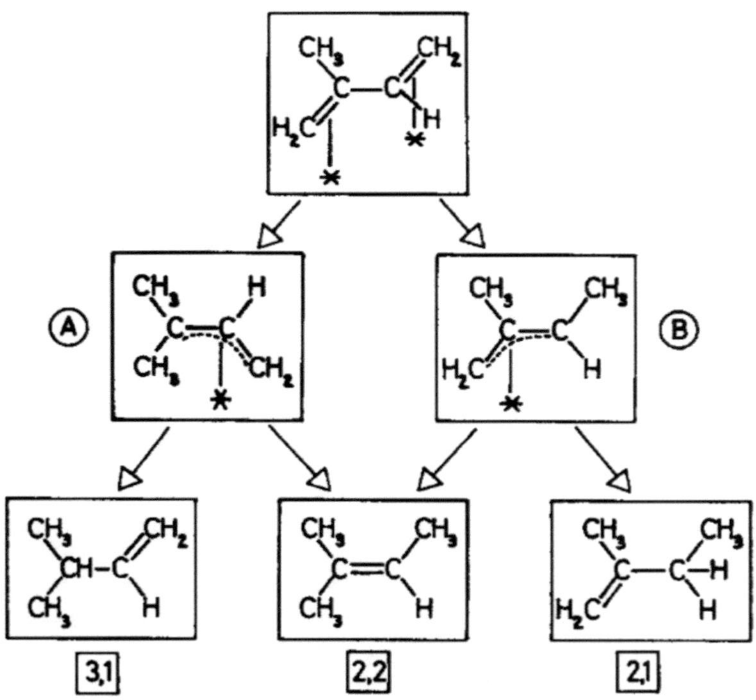

Figure 11.
Reduction pathway for isoprene to form monoene product. Reproduced from [6] with permission from the American Chemical Society.

Ornelas et al. also studied the semi-heterogeneous catalysis by using dendrimer-passivated palladium nanoparticle as a catalyst for the hydrogenation of dienes [28]. They synthesized the 1,2,3-triazole heterocycles-capped palladium nanoparticle catalyst that exhibits the higher reactivity to diene hydrogenation compared to the

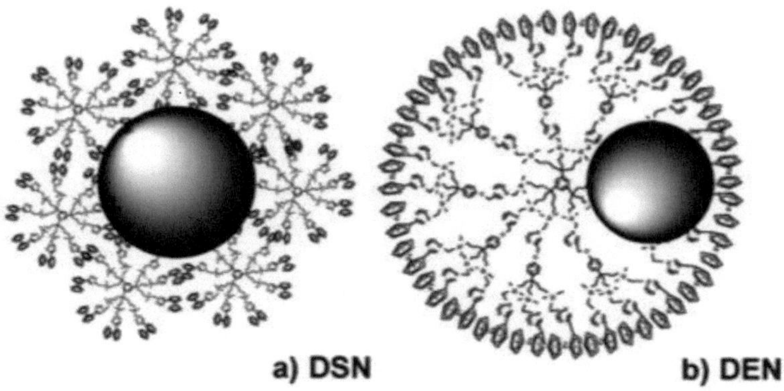

Figure 12.
(a) Generation 1, dsn, and (b) generation 1, den, Pd nanoparticle encapsulated by PAMAM. Reproduced from [28] with permission from the Royal Society of Chemistry.

PAMAM dendrimer-encapsulated Pd nanoparticle catalyst. Depending on the generation of the triazole dendrimer, the catalyst could develop into different morphologies that control the activity and selectivity. When the G_0 dendrimers are used, the interdendrimer-stabilized palladium nanoparticle (DSN) is formed (**Figure 12**). For the higher generation dendrimers, the intradendrimer-encapsulated palladium nanoparticle (DEN) is produced. Due to the smaller size of the G_0 dendrimer, the Pd nanoparticle cannot be encapsulated by dendrimer and needs to be stabilized by several G_0 dendrimer. This makes the overall size of DSN relatively larger than the other higher generation dendrimer-capped catalysts.

When the higher generation dendrimer is used, the large size of dendrimer allows enough Pd ions to be encapsulated in the interior of dendrimer and the following reduction generates dendrimer-encapsulated Pd nanoparticles. Unlike PAMAM-stabilized metal nanoparticle, 1,2,3-triazoleferrocenyl dendrimer-stabilized Pd nanoparticle would have higher stability during the catalyst reaction [28]. The nature of the reducing agent and the generation of dendrimer are found to have noticeable influence on the reactivity and stability of each catalyst. DEN-G_1 reduced by methanol has the best reactivity for the mono-hydrogenation of diene, which indicates that the smaller size of the Pd nanoparticle increases the reactivity. Moreover, the unique structure of 1,2,3-triazoleferrocenyl dendrimer is also found to be the another reason for catalyst to have higher reactivity. **Table 9** shows DEN-G_1 has higher reactivity to small dienes for mono-hydrogenation. However, the catalytic reactions of large dienes with steric bulkiness are slightly slower. Hydrogenation of trienes mostly results in the formation of monoene compounds indicating the high activity of diene intermediate after mono-hydrogenation. More substituted dienes tend to have a lower catalytic reactivity. Moreover, the isomerization of terminal monoenes and the trace amount of the 1,4-hydrogenation product from highly substituted dienes are also observed in the reaction.

3.3 Well-defined small organic ligand-capped palladium nanoparticles as semi-heterogeneous catalysts

Many ligand-passivated nanoparticles have been used as semi-heterogeneous catalysts. Since the surface ligands that stabilize nanoparticles from aggregation can have either hydrophobic or hydrophilic property, they can be soluble in various solvents including organic and aqueous solutions. Shon group has developed the thiosulfate protocol using alkanethiosulfate as a ligand precursor to passivate and

Substrate	Product (yield)	TOF[a]
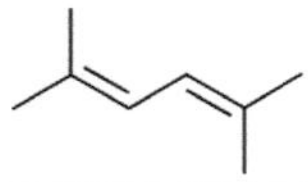	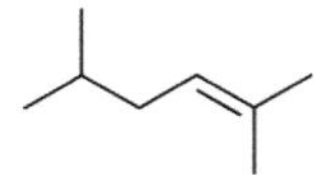 (78%)	810
	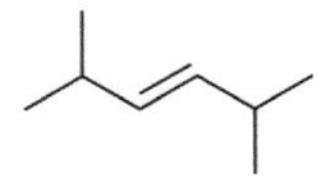 (22%)	230
	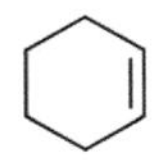 (100%)	1150
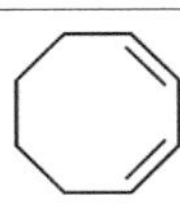	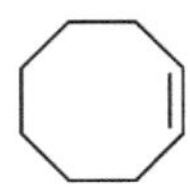 (100%)	530

[a]TOF were determined by the yield of formation and final product.
Data reproduced from [28].

Table 9.
Hydrogenation of olefins catalyzed by methanol reduced DEN-G_1 (Reaction condition: 25°C, 1 atm H_2, and substrate/Pd ratio is 1000 in mixed solvent $CHCl_3$/MeOH = 2).

stabilize the catalytically active palladium nanoparticle surface [29–33]. The hydrophobic alkanethiolate ligand gives the nanoparticle high solubility in nonpolar organic solvent. Since alkanethiosulfate offers slower passivation activity, the surface ligand density of alkanethiolate on Pd nanoparticles can be controlled. Therefore, the alkanethiolate-capped Pd nanoparticles generated from alkylthiosulfate exhibit good catalytic activity and selectivity toward various organic reactions including isomerization and hydrogenation. Hexanethiolate- and dodecanethiolate-capped Pd nanoparticles show unique catalytic properties for the reaction of allyl alcohol under the atmospheric pressure of hydrogen gas at room temperature [30, 31]. Allyl alcohol can undergo either hydrogenation to 1-propanol or isomerization to propanal.

Octanethiolate-capped Pd nanoparticle (C8 PdNP) with average core size of ~2.3 nm is synthesized and its composition and structure are confirmed by various instruments. Alkanethiolate-capped Pd nanoparticles are investigated for the catalytic reaction of 1,4-pentadiene [32]. The mechanistic studies show that mono-hydrogenation of isolated dienes would take place on one of the terminal C=C bond (**Figure 13**). The di-σ-bonded Pd intermediate **A** would form after the di-π-bond adsorption on Pd, which leads to the mono-hydrogenation and the formation of 1-pentene. The further isomerization via mono-σ-bonded Pd intermediate converts 1-pentene to the isomerized product, 2-pentene.

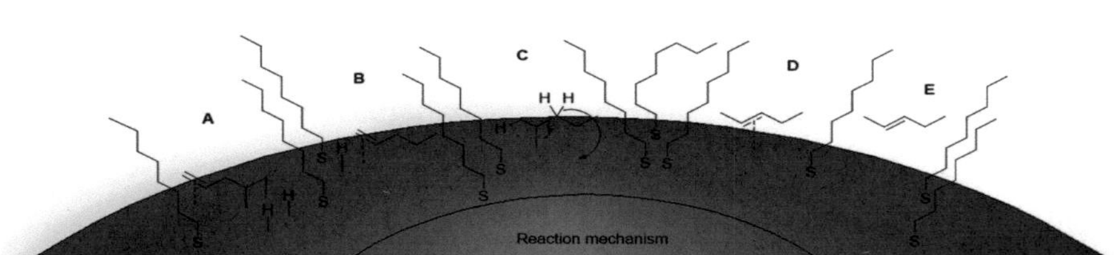

Figure 13.
The proposed mechanism (from A to E) for 1,4-pentadiene catalytic reaction under H_2 environment. Reproduced from [32] with permission from the Royal Society of Chemistry.

The C8 PdNP is used as the catalyst for the hydrogenation of various conjugated diene and triene substrates as shown in **Table 10** [33]. The catalytic reactions of conjugated dienes with different substitution patterns around C=C bonds produce almost exclusively the mono-hydrogenation products (**Table 10**, Entries 1–5). In addition, the hydrogenation of trienes (Entries 6 and 7) also results in the high yields of isolated dienes, the mono-hydrogenation products. This reactivity confirms the important role of alkanethiolate ligands on controlling the activity of Pd catalyst surface. With its surface passivated by organic ligands, nanoparticle could maintain a good stability, so that it could avoid aggregation and keep the large surface area intact. The surface ligand could block the more active site (terrace surface) for hydrogenation and promotes selective hydrogenation of dienes. The analysis of final monoene compositions showed that the major product is the 1,4-addition product and the minor product is the 1,2-addition product. The kinetic study of diene to monoene proved that the high selectivity for the 1,4-addition product is the result of both initial 1,4-addition reaction and the subsequent isomerization of terminal alkene, the 1,2-addition product, into internal alkene.

After the first hour of reaction, the conversion of diene in entry 1 reaches over 50% with the ratio of 1,4-/1,2-addition products at 3.43. The consumption of reactant is almost complete after 5 h reaction with the ratio of 1,4-/1,2-addition products at 4.90. The yield of the 1,4-addition product is continuously increasing after 5 h reaction with the ratio of 1,4-/1,2-addition products increasing to 10.30 at the 24 h reaction. This clearly indicated that the isomerization of the terminal alkene to the internal alkene is the reason for the high selectivity of 1,4-addition product. The selectivity between the 1,4- and 1,2-addition products are also summarized for other diene and triene substrates as shown in **Table 10**. C8 PdNP clearly showed high selectivity for the 1,4-addition product, the thermodynamically more stable product. C8 PdNP clearly exhibits excellent selectivity to form internal alkene, the mono-hydrogenation and 1,4-addition product, compared to other reported catalytic systems. Not only the conversion yields and selectivity are superior but also the reaction condition (room temperature and atmospheric pressure) is much

Entry	Substrate	Reaction condition	Major product (%)
1	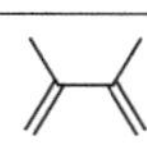	C8 PdNP 24 h	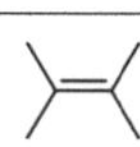 (91% 1,4-) + (9% 1,2-)
2	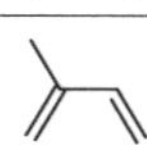	C8 PdNP 24 h	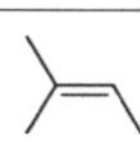 (93% 1,4-) + (7% 1,2-)
3	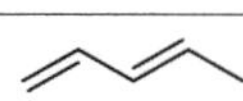	C8 PdNP 24 h	 (92% 1,4-) + (8% 1,2-)
4	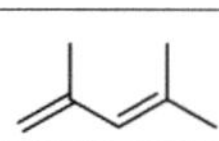	C8 PdNP 24 h	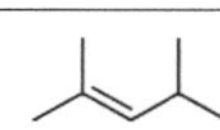 (90% 1,4-) + (7% 1,2-)
5	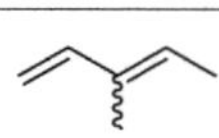	C8 PdNP 24 h	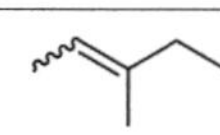 (90% 1,4-) + (5% 1,2-)
6	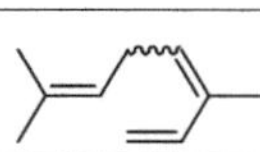	C8 PdNP 24 h	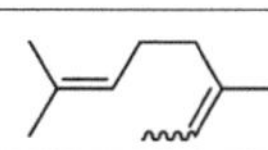 (59% 1,4-) + (41% 1,2-)
7	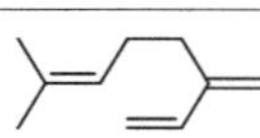	C8 PdNP 24 h	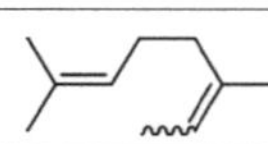 (69% 1,4-) + (23% 1,2-)

Data reproduced from [29].

Table 10.
Catalysis results of various dienes and trienes with 5 mol% of octanethiolate-capped Pd nanoparticle in $CDCl_3$ at 1 atm H_2 after 24 h.

friendlier than other homogenous and heterogeneous catalysts tested for diene hydrogenation.

4. Conclusions

This chapter reviewed the use of various catalysts including metal complexes, supported metals and metal nanoparticles, colloidal nanoparticles for selective hydrogenation of polyunsaturated olefins, mostly dienes and trienes. Selective hydrogenation of polyenes has been considered as an important process in many chemical and pharmaceutical industries. By controlling chemical environments around active catalytic sites using organic ligands, inorganic dopants, ionic liquids, dendrimers, secondary metals, etc., the catalytic activity and selectivity of various catalysts for partial hydrogenation of polyenes could be improved. Especially, nanocatalyzed selective hydrogenation represents a rapidly growing field, but there is still much work to be done to generate industrially viable nanocatalysts that can be operated under many catalytic cycles with acceptable integrity. In addition, more in-depth understanding of critical structure–function relationships should be obtained for the development of optimized nanocatalysts with high chemoselectivity and stereoselectivity.

Acknowledgements

This research was funded by National Institute of General Medical Science [GM089562].

Conflict of interest

The authors declare no conflict of interest. The funding sponsor had no role in the design of the study, in the writing of the manuscript, and in the decision to publish the research results.

Author details

Ting-An Chen and Young-Seok Shon*
Department of Chemistry and Biochemistry, California State University, Long Beach, CA, USA

*Address all correspondence to: ys.shon@csulb.edu

References

[1] Bond GC. Metal-Catalysed Reactions of Hydrocarbons. New York: Springer; 2005

[2] Derrien ML. Studies in Surface Science and Catalysis. 1986;**27**:613

[3] Erman WF. Chemistry of the Monoterpenes: An Encyclopedic Handbook. New York, NY: M. Dekker; 1985. p. 1709

[4] Mimoun H. Catalytic opportunities in the flavor and fragrance industry. CHIMIA International Journal for Chemistry. 1996;**50**(12):620-625

[5] Raoult Y, Choukroun R, Basso-Bert M, Gervais D. Hydrogenation and isomerization of olefins with diphenylphosphinomethyl hydride zirconium, $[Cp_2ZrH(CH_2PPh_2)]_n$, a selective homogeneous catalyst. Journal of Molecular Catalysis. 1992;**72**(1):47-58

[6] Bond GC, Rawle AF. Catalytic hydrogenation in the liquid phase. Part 1. Hydrogenation of isoprene catalysed by palladium, palladium-gold and palladium-silver catalysts. Journal of Molecular Catalysis A: Chemical. 1996;**109**(3):261-271

[7] Schrock RR, Osborn JA. Catalytic hydrogenation using cationic rhodium complexes. 3. The selective hydrogenation of dienes to monoenes. Journal of the American Chemical Society. 1976;**98**(15):4450-4455

[8] Frankel EN, Butterfield RO. Homogeneous hydrogenation of diolefins catalyzed by tricarbonyl chromium complexes. I. Stereoselective 1,4-addition of hydrogen. The Journal of Organic Chemistry. 1969;**34**(12): 3930-3936

[9] Valentine D Jr, Johnson KK, Priester W, Sun RC, Toth K, Saucy G. Rhodium chiral monophosphine complex catalyzed hydrogenations of terpenic and .alpha.-(acylamino)-substituted acrylic acids. The Journal of Organic Chemistry. 1980;**45**:3698

[10] Panella L, Feringa BL, de Vries JG, Minnaard AJ. Enantioselective Rh-catalyzed hydrogenation of enol acetates and enol carbamates with monodentate phosphoramidites. Organic Letters. 2005;**7**:4177

[11] Takaya H, Ohta T, Sayo N, Kumobayashi H, Akutagawa S, Inoue S, et al. Enantioselective hydrogenation of allylic and homoallylic alcohols. Journal of the American Chemical Society. 1987; **109**:1596-1598

[12] Liu J, Krajangsri S, Singh T, de Seriis G, Chumnanvej N, Wu H, et al. Regioselective iridium-catalyzed asymmetric monohydrogenation of 1,4-dienes. Journal of the American Chemical Society. 2017;**139**: 14470-14475

[13] Chang H, Phillips J. Catalytic synergism in physical mixtures of supported iron-cerium and supported noble metal for hydroisomerization of 1,3-butadiene. Langmuir. 1997;**13**: 477-482

[14] Sales EA, de Jesus Mendes M, Bozon-Verduraz F. Liquid-phase selective hydrogenation of hexa-1,5-diene and hexa-1,3-diene on palladium catalysts. Effect of tin and silver addition. Journal of Catalysis. 2000;**195**: 96-105

[15] Xu W, Ni J, Zhang Q, Feng F, Xiang Y, Li X. Tailoring supported palladium sulfide catalysts through H_2-assisted sulfidation with H_2S. Journal of Materials Chemistry A. 2013;**1**: 12811-12817

[16] Bachiller-Baeza B, Iglesias-Juez A, Castillejos-López E, Cuerrero-Ruiz A,

Di Michiel M, Fernández-García M, et al. Detecting the genesis of a high-performance carbon-supported Pd sulfide nanophase and its evolution in the hydrogenation of butadiene. ACS Catalysis. 2015;**5**:5235-5241

[17] Kahsar KR, Schwartz DK, Medlin JW. Selective hydrogenation of polyunsaturated fatty acids using alkanethiol self-assembled monolayer-coated Pd/Al_2O_3 catalysts. ACS Catalysis. 2013;**3**:2041-2044

[18] Karakhanov E, Maximov A, Kardasheva Y, Semernina V, Zolotukhina A, Ivanov A, et al. Pd nanoparticles in dendrimers immobilized on silica–polyamine composites as catalysts for selective hydrogenation. ACS Applied Materials & Interfaces. 2014;**6**(11):8807-8816

[19] Yi H, Du H, Hu Y, Yan H, Jiang HL, Lu J. Precisely controlled porous alumina overcoating on Pd catalyst by atomic layer deposition: Enhanced selectivity and durability in hydrogenation of 1,3-butadiene. ACS Catalysis. 2015;**5**:2735-2739

[20] Bemsmeier D, Chuenchom L, Paul B, Rummler S, Smarsly B, Kraehnert R. Highly active binder-free catalytic coatings for heterogeneous catalysis and electrocatalysis: Pd on mesoporous carbon and its application in budadiene hydrogenation and hydrogen evolution. ACS Catalysis. 2016;**6**:8255-8263

[21] Hugon A, Delannoy L, Krafft JM, Louis C. Selective hydrogenation of 1,3-butadiene in the presence of an excess of alkenes over supported bimetallic gold-palladium catalysts. Journal of Physical Chemistry C. 2010;**114**:10823-10835

[22] Zweni PP, Alper H. Silica-supported dendrimer-palladium complex-catalyzed selective hydrogenation of dienes to monoolefins. Advanced Synthesis and Catalysis. 2006;**348**(6): 725-731

[23] Yardimci D, Serna P, Gates BC. Tuning catalytic selectivity: Zeolite- and magnesium oxide-supported molecular rhodium catalysts for hydrogenation of 1,3-butadiene. ACS Catalysis. 2012;**2**: 2100-2113

[24] Speziali MG, Moura FCC, Robles-Dutenhefner PA, Araujo MH, Gusevskaya EV, dos Santos EN. Selective hydrogenation of myrcene catalyzed by complexes of ruthenium, chromium, iridium and rhodium. Journal of Molecular Catalysis A: Chemical. 2005;**239**(1–2):10-14

[25] Robles-Dutenhefner PA, Speziali MG, Sousa EMB, dos Santos EN, Gusevskaya EV. Selective hydrogenation of myrcene catalyzed by sol–gel Pd/SiO_2. Applied Catalysis A: General. 2005;**295**(1):52-58

[26] Luza L, Rambor CP, Gual A, Bernardi F, Domingos JB, Grehl T, et al. Catalytically active membrane like devices: Ionic liquid hybrid organosilicas decorated with palladium nanoparticles. ACS Catalysis. 2016;**6**(10):6478-6486

[27] Niu Y, Crooks RM. Preparation of dendrimer-encapsulated metal nanoparticles using organic solvents. Chemistry of Materials. 2003;**15**(18): 3463-3467

[28] Ornelas C, Salmon L, Aranzaes JR, Astruc D. Catalytically efficient palladium nanoparticles stabilized by "click" ferrocenyl dendrimers. Chemical Communications. 2007;**46**:4946-4948

[29] San KA, Shon Y-S. Synthesis of alkanethiolate-capped metal nanoparticles using alkyl thiosulfate ligand precursors: A method to generate promising reagents for selective catalysis. Nanomaterials. 2018;**8**:346. DOI: 10.3390/nano8050346

[30] Sadeghmoghaddam E, Gu H, Shon YS. Pd nanoparticle-catalyzed isomerization vs hydrogenation of allyl alcohol: Solvent-dependent regioselectivity. ACS Catalysis. 2012; **2**(9):1838-1845

[31] Gavia DJ, Shon Y-S. Controlling surface ligand density and core size of alkanethiolate-capped Pd nanoparticles and their effects on catalysis. Langmuir. 2012;**28**(40):14502-14508

[32] Zhu JS, Shon Y-S. Mechanistic interpretation of selective catalytic hydrogenation and isomerization of alkenes and dienes by ligand deactivated Pd nanoparticles. Nanoscale. 2015;**7**(42): 17786-17790

[33] Chen T-A, Shon Y-S. Alkanethioate-capped palladium nanoparticle for selective catalytic hydrogenation of dienes and trienes. Catalysis Science & Technology. 2017;7:4823-4829. DOI: 10.1039/C7CY01880K

Chapter 7

Tailoring the Kinetic Behavior of Hydride Forming Materials for Hydrogen Storage

Julián Atilio Puszkiel

Abstract

Hydride forming materials, i.e., binary, complex hydrides, and their mixtures, have been extensively investigated owing to their potential hydrogen storage properties. They possess high volumetric hydrogen capacity and relative high gravimetric hydrogen capacity. However, one of the main constraints for their practical application is their slow kinetic behavior. For this reason, enormous effort has been devoted to improve the hydrogenation and dehydrogenation rates. Several strategies have been developed for the enhancement of the kinetic behavior of the most relevant hydride forming materials such as MgH_2, MBH_4 (M = Li, Ca, Mg, Na, K), MNH_2 (M = Li and Mg), MBH_4 + 'MH_2 (M = Li, Ca, Mg; 'M = Li, Mg, Ca), and MNH_2 + 'MH_2 (M = Li, Mg; 'M = Li). Tuning the kinetic behavior of these hydride forming materials involves different approaches and their combinations. The most relevant approaches are: (1) improving the microstructural refinement *via* mechanical milling, (2) doping with transition metal and transition metal compounds, (3) forming *in situ* catalyst, and (4) nanoconfining doped hydride forming materials. Herein, basic concepts about the chemical reaction for the hydride compound formation/decomposition, thermodynamics, kinetics, and applied strategies to enhance the kinetic behavior of hydride compounds and systems are comprehensively described and discussed.

Keywords: kinetic behavior, hydrogen storage, hydrides, catalyst, modeling

1. Introduction

1.1 Hydrogen as the future renewable energy vector

In the last two centuries, the energy supply has been based on fossil fuels (natural gas, liquid fuels, and coal). This strong dependence of our modern society on fossil fuels has economic- and geo-political consequences and causes detrimental environmental footprints. **Figure 1A** shows the world fossil fuel demand-supply trend from the 1990s until 2017. In 2017, the world fossil fuel demand (97.8 barrel per day) is over the supply (97.4 barrel per day) [1]. For these reasons, in the last decades, many efforts have been put for the development and research of energy renewable sources. The market of the renewable energies achieved a new global record in 2015 with $285 billion investment on this technology, exceeding the previous record of $278 billion on 2011 [2]. The use of renewable energy has many

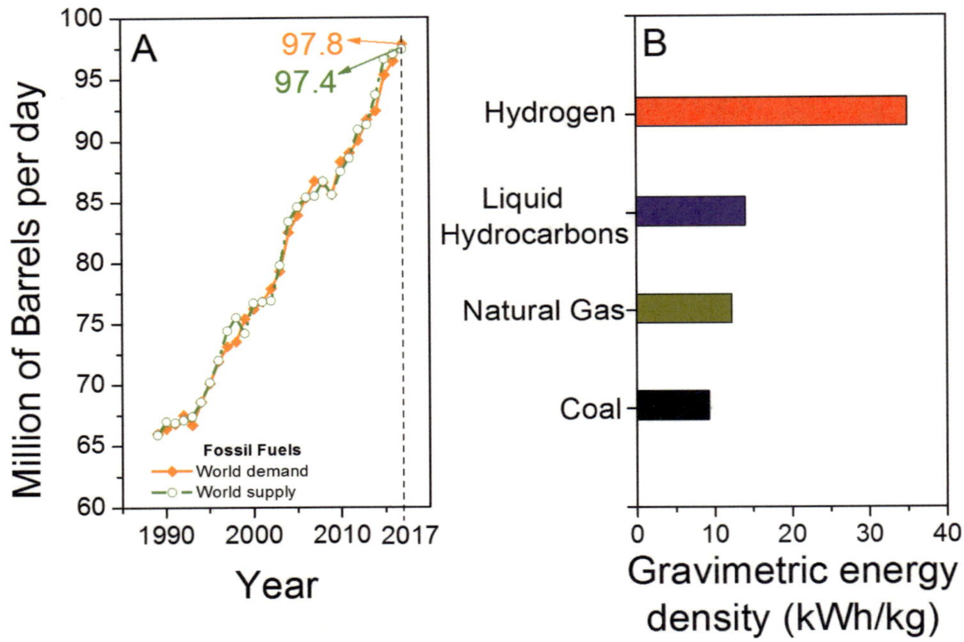

Figure 1.
(A) Evolution of the world demand and supply of fossil fuels in millions of barrels per day (includes all kinds of fossil fuels) [1] and (B) Gravimetric energy density for pure: hydrogen, liquid hydrocarbons, natural gas and coal [3].

potential benefits, including a reduction in CO_2 emissions, the diversification of energy vectors, and a reduced dependency on fossil fuel markets. In this regard, hydrogen (H_2) is considered as a potential energy vector because it is abundant and its combustion generates a large amount of energy and water: $H_2 + ½ O_2 \rightarrow H_2O$ + Energy. **Figure 1B** shows that H_2 exhibits far higher gravimetric energy density than other traditional fossil fuels: 33.3 kWh/kg for H_2, 13.9 kWh/kg for natural gas, and 12.4 kWh/kg for liquid hydrocarbons [3–6]. Furthermore, H_2 can be directly produced from clean sources such as, for instance, from electrolysis of water or from biomass with light from the sun and biological micro-organism. These renewable methods for hydrogen production can avoid the coal reforming, though this technology is under development and it is not available on industrial scale yet [7].

Nowadays, H_2 is used as feedstock in the petrochemical and fertilizers industries. There are restricted technological applications of H_2 as for example in portable devices like mobile phones, computers, and even some hydrogen driven cars. In spite of all the advantages of H_2 to be considered as the new energy carrier and the evolution of the hydrogen technology showing some technological applications, there are some major issues to be solved before introducing the "hydrogen technology" into the massive market [8]. The hydrogen technology is still under early research and development phase, the current costs of this new technology and the lack of established infrastructure are the main constraints [9–11]. Among the technological problems for the implementation of the "hydrogen economy," one of the main bottleneck is the lack of safe, compact, energetic-, and cost-efficient hydrogen storage system, especially for mobile applications such as means of transports [12, 13].

On one hand, H_2 has the highest gravimetric energy density in comparison with the traditional fossil fuels, **Figure 1B**. On the other hand, H_2 has extremely low density at room temperature (0.089 g/dm^3 under 1 bar), hence its volumetric energy density is extremely low (0.003 kWh/dm^3). **Figure 2A** shows that the volumetric energy density of H_2 is notably lower than the traditional fossil fuels (0.008 kWh/dm^3 for natural gas, 10.5 kWh/dm^3 for liquid hydrocarbons, and 35 kWh/dm^3 for processed coal). **Figure 2B** shows the volumetric storage density for different traditional and hydrogen storage systems. As seen, there is a huge gap between the volumetric energy density of the light and cheap tank system for liquid hydrocarbons such as gasoline and the volumetric energy density of hydrogen storage systems.

Liquefaction and compression are the traditional physical methods to store hydrogen. Nowadays, available liquid hydrogen storage systems can achieve energy densities of 1.2 kWh/dm^3. However, cryogenic hydrogen storage systems work at −253°C (liquefied hydrogen temperature) and this requires 20–30% of its energy content. Moreover, hydrogen losses because of evaporation are between 0.3 and 3% per day. On the contrary, gas storage systems can confine hydrogen for long periods without losses. The gas hydrogen vessels need similar technology and infrastructure to that of the established compressed natural gas tanks. Reinforced gas hydrogen containers for 350 and 700 bar of pressure are commercially available. These gas storage systems can achieve volumetric energy densities up to 0.9 kWh/dm^3, but about 15% of the fuel energy content is required for compression. Solid storage systems in hydride compounds are so far tested on small and middle scale in laboratories. As an example, a tested solid storage system has lower volumetric energy density (0.8 kWh/dm^3) than the traditional physical storage systems [14]. However, solid storage systems work at milder temperature and pressure conditions than the physical storage systems, i.e., in the range of temperatures between 25 and 400°C and under pressure in the range of 10–100 bar of H_2, depending on the hydride material. This means that extreme cryogenic temperature of −253°C or extreme high pressure of 700 bar is not necessary. It brings two main advantages from the safety and energy point of view: (1) mechanical properties of the vessel's material are not demanding as for gas storage containers, (2) energy losses because of hydrogen liquefaction, compression, or boil-off are avoided. Nonetheless, the volumetric energy density of the solid storage hydrogen system is still low, particularly comparing with the gasoline storage system. Moreover, there are critical topics like materials' properties (hydride material), charging and discharging conditions of the systems (pressure, time, and heat transfer), charging

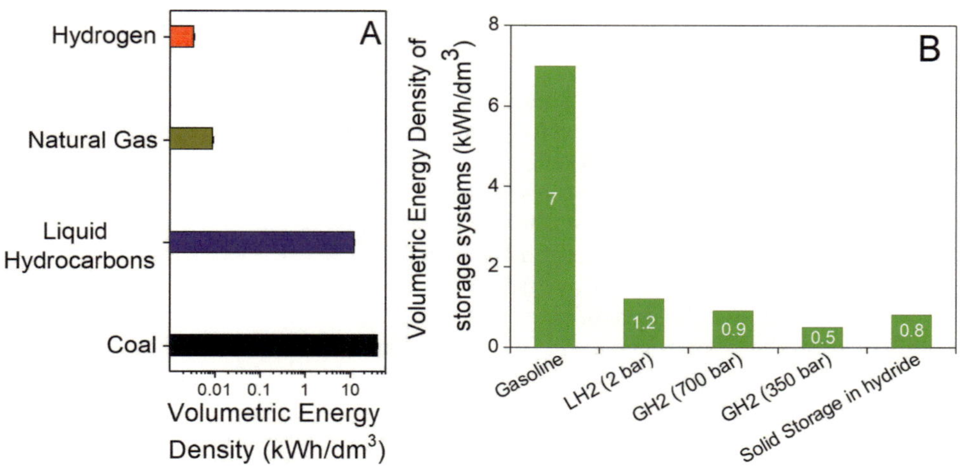

Figure 2.
(A) Volumetric energy density for pure: hydrogen, liquid hydrocarbons, natural gas, and coal [3].
(B) Volumetric energy density for storage systems [14].

cycles, and cost that are still matter of intensive research. Improving mainly the properties of the hydride forming material and optimizing the weight and operative conditions of the solid hydrogen storage system is challenging. However, there are several promising materials under research and development, and besides unexplored system configurations, which can lead to an efficient and safe hydrogen storage system for the future hydrogen economy.

2. Hydride compounds: basics on the chemical reaction and thermodynamics

The solid storage system is based on hydride forming materials, i.e., the system consists basically in a vessel filled with the hydride forming material. Of course, the complexity of the storage system including pipes, heat transfer system, and control instrumentations is in this simplified description neglected. In this section, it is presented basic concepts regarding the chemical reaction for the hydride formation and its thermodynamics.

2.1 Reversible gas: solid chemical reaction

Hydrogen interacts with a large number of elements and materials for the formation of hydride compounds. Hydride forming materials can be classified as: metals (alkaline, alkaline earth, transition, and rare earth), intermetallic (Mg_2Ni, $LaNi_5$, etc.), non-intermetallic (Mg-Fe, Mg-Co, etc.), and the combination of boron (B), aluminum (Al), or nitrogen (N) with alkaline or alkaline earth metals. Furthermore, a general classification considering the main feature of different hydride compounds is shown in **Table 1**. There are two main groups of hydride: the room temperature hydrides where hydrogen is located in the interstices of the metal's lattice, without forming a strong metal-hydrogen bond. This kind of hydride works at low temperatures, i.e., 20–50°C, and under low and high pressures, depending on the type of alloy. For instance, $LaNi_5$ alloy works at low pressure, between 10 and 50 bar, while TiCrMo alloy uptake hydrogen under high pressures, over 100 bar. The main constraint of the room temperature hydride compounds is the low gravimetric hydrogen storage capacities ranging from 1 to 3 wt.% H_2, though they do have considerable hydrogen volumetric capacity of about 100 kg/m^3. The other categories are binary and complex hydrides, where hydrogen is chemically bound. They work in a broad range of temperatures and pressures, but the most common ranges are temperatures over 100°C and pressures from 10 to 200 bar. There are some exceptions such as non-stable hydride compounds at room temperature, as for example $Ti(BH_4)_3$, $Fe(BH_4)_2$, $Ni(AlH_4)_2$, among others. Moreover, several binary and complex hydrides are not even reversible [15–20]. These hydrides show large gravimetric hydrogen storage capacities ranging from 4 to about 20 wt.% H_2 and also considerable volumetric hydrogen capacity from 100 to 150 kg/m^3.

The chemical reaction for the hydride formation can be described as a reversible gas-solid reaction. For the sake of clarity, the most simple reaction between a solid metal ($M_{(s)}$) and gas hydrogen ($H_{2(g)}$) is herein explained. At certain temperature (T) and under certain pressure (P), $M_{(s)}$ reacts with $H_{2(g)}$ to form a hydride compound ($MH_{x(s)}$) according to reaction (1):

$$M_{(s)} + (X/2)H_{2(g)} \leftrightarrows MH_{x(s)} + \text{Heat}\ (T, P) \tag{1}$$

Reaction (1) shows the overall process for the reversible formation of a hydride compound without any detail about reaction intermediates. As seen, the hydride

Hydrogen bond	Hydride compounds	Examples of hydride forming materials (hydride)	Main characteristics
Interstitial hydrogen	Room temperature hydrides	V (VH_2), $LaNi_5$ alloy ($LaNi_5H_6$), FeTi alloy ($FeTiH_2$), TiCrMo, TiVMn alloys, etc.	Form AyBx (y = 2 and x = 1, 2, 5); A= early transition or rare-earth metal, B= trandition metal , rare-earth or alkaline metal.
Chemically bound hydrogen	Binary hydrides composed of Alkaline and alkaline rare earth	Li (LiH), Mg (MgH_2), Na (NaH), etc.	Hydrogen is chemically bound (ionic, covalent or a mixture of both).
	Complex hydrides composed of transition metals	Mg-Fe (Mg_2FeH_6), Mg_2Ni (Mg_2NiH_4), etc.	$Mg2^+[THn]^{\delta-}$ (T= Mi, Co, Fe, among others, n= 4, 5 and 6).
	Complex hydrides containing boron (B): Borohydrides	Li-B ($LiBH_4$), Ca-B ($Ca(BH_4)_2$), etc.	$[BH_4]^-M^+$ (M= alkaline, earth - alkaline, transition, rare-earth metal).
	Complex hydrides containing aluminum (Al): Alanates	Li-Al ($LiAlH_4$), Na-Al ($NaAlH_4$), etc.	$[AlH_4]^-M^+$ (M= alkaline, earth-alkaline, transition, rare-earth metal).
	Complex hydrides containing nitrogen (N): Amides	Li-N ($LiNH_2$), $Mg(NH_2)_2$, etc.	$[NH_2]^-M^+$ (M= alkaline, earth-alkaline, transition, rare-earth metal).

Table 1.
General classification of hydride compounds and their main characteristics [15–20].

compound formation is exothermic, while its decomposition is endothermic. Moreover, the formation and decomposition of a hydride compound occur under certain T and P conditions, which depend on the kind of $M_{(s)}$ and the resulting hydride compound. Thus, these T and P conditions are determined by the thermodynamics and kinetics of the metal-hydrogen ($M\text{-}H_2$) system.

2.2 Thermodynamics

Thermodynamic properties for the $M\text{-}H_2$ systems are usually characterized by measuring pressure-composition-isotherms (PCIs). **Figure 3A** shows ideal PCIs (without slope and hysteresis), where the x-axis is the hydrogen concentration (C_H) expressed as the ratio between atomic hydrogen and metal (H/M), and y-axis is the hydrogen pressure (pH_2). The procedure to measure a PCI consist in introducing the hydride forming metal or material in a sealed vessel connected to hydrogen supply and increase steeply the hydrogen pressure at a constant temperature. There are different steps involved during the hydrogen absorption process in a metal under equilibrium conditions. For example, taking the PCI at T_2 (**Figure 3A**), a detailed description of the process during PCI characterization can be done as follows [4, 21]:

1. At the beginning of the process, molecular hydrogen (H_2) near the metal's surface suffers van der Waals interactions. This H_2-M interaction process is called physisorption.

2. Molecular H_2 dissociates on the metal's surface ($H_2 \rightarrow H + H$) and forms a M-H bond. This process is called chemisorption and the required energy for it depends on the elements on the metal's surface.

3. Chemisorbed H diffuses to the interstitial site of the metal's lattice and exothermically dissolves to form a solid solution (α phase). This process occurs in the low hydrogen concentration zone (H/M < 0.1), as shown in **Figure 3A**—point (a) and is described by reaction (2):

$$M_{(s)} + (Y/2)\, H_{2(g)} \leftrightarrows MH_{Y(s)}\ (\alpha\ \text{phase}) \quad (2)$$

4. Increasing the hydrogen pressure, the solid solution (α phase) reaches a hydrogen saturation concentration at H/M = 0.1. From this hydrogen concentration, the hydride phase (β phase) starts to form as shown in **Figure 3A**—point (b). Significant expansions of the metal's lattice causes strong H–H interactions which result in the nucleation and then further growth of the hydride phase. In an ideal case, the overall process of formation of the hydride phase occurs under constant pressure and is called plateau region in the PCI curve, **Figure 3A**—from point (b) to point (d). In this region, the solid solution (α phase) and the hydride phase (β phase) co-exist in equilibrium conditions, **Figure 3A**—point (c). The pressure associated to the plateau region is called equilibrium pressure ($P_{eq.}$). Reaction (3) describes the process:

$$MH_{Y(s)}\ (\alpha\ \text{phase}) + [(X\text{–}Y)/2]\, H_{2(g)} \leftrightarrows MH_{x(s)}\ (\beta\ \text{phase}) \quad (3)$$

5. Once the formation of the hydride phase is finished, the hydrogen pressure increases again, as shown in **Figure 3A**—from point (d). This phenomenon is ascribed to the atomic hydrogen dissolution in the hydride phase.

Under experimental conditions, the equilibrium pressure during the "plateau region" is not perfectly constant, as shown in **Figure 3C**. The sloppy plateau is attributed to expansions of the lattice of the hydride and relaxations of the metallic matrix, which causes a slight increase of the equilibrium pressure during the chemical reaction of the hydride phase (β phase) formation. Furthermore, it is also observed in

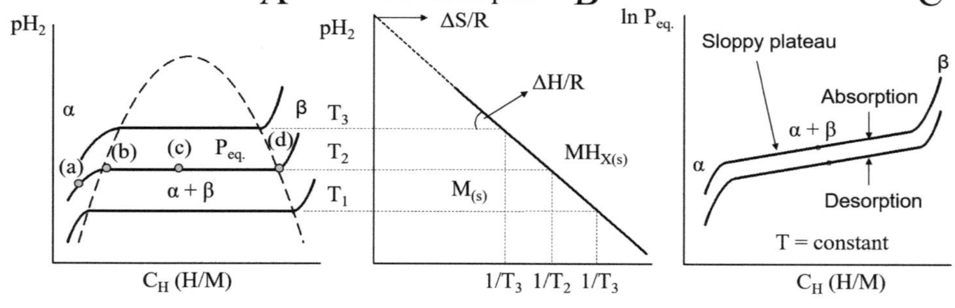

Figure 3.
(A) Ideal PCIs for the description of the hydrogen absorption-desorption process through hydride compound formation in equilibrium conditions. (B) van't Hoff plot built from PCIs. (C) PCIs showing the real behavior with slope and hysteresis.

Figure 3C that the equilibrium pressure for the hydrogenation process is lower than the equilibrium pressure for the dehydrogenation process. This phenomenon is called hysteresis and it is ascribed to localized defects in the metallic lattice and inhomogeneity on the metal's surface [21–23]. Therefore, different equilibrium pressures for the hydrogenation and dehydrogenation processes are usually reported.

The dependence of the equilibrium pressure ($P_{eq.}$) on the temperature (T) is given by the van't Hoff Eq. (4):

$$ln\, P_{eq} = \left[\frac{\Delta H}{RT}\right] - \left[\frac{\Delta S}{R}\right], \tag{4}$$

where ΔH and ΔS are the enthalpy and entropy changes for the chemical reaction of the hydride phase formation, respectively, and R is the ideal gas constant. Values for ΔH and ΔS can be obtained from the slope and intercept of the linear correlation between the $\ln(P_{eq})$ and 1/T, as shown in **Figure 3B**. This linear correlation is built from the equilibrium pressures determined by the PCI measurements (pH_2 during the plateau region—**Figure 3A**—from point (b) to point (d)) at each temperature (T_1, T_2, etc.). ΔS indicates the entropy change for hydrogen, i.e., from molecular gas hydrogen to hydrogen in the hydride phase. For metal-hydrogen systems, the standard entropy change for hydrogen is 130 kJ/K mol, but it can have a different value for other kind of hydride systems as for example for hydrides composed of boron, aluminum, or nitrogen and alkaline or alkaline earth metals [4, 21]. ΔH of the hydride compound or system characterizes the stability of the metal-hydrogen bond (M–H) and it takes a negative value for the exothermic hydrogenation and a positive value for the endothermic dehydrogenation. These two thermodynamic parameters, i.e., ΔH and ΔS, are quite important for hydride forming material design and hence for practical applications. They allow calculating the temperature for the hydrogen release from hydride phase under atmospheric pressure (~1 bar), which refers to the minimum temperature for the dehydrogenation process without any kind of kinetic restriction. Additionally, the thermodynamic properties of the hydride phase provide information about the range of temperature and pressures at which the hydride system works. **Table 2** shows the experimental enthalpy, entropy values, and decomposition temperature under 1 bar

Hydride compound/ Hydride system	ΔH [kJ/mol H_2]	ΔS [J/K mol H_2]	$T_{decomposition}$ = ΔH/ΔS at 1bar [°C]	Ref.
$LaNi_5H_6$	31.8	110.0	16	[24]
MgH_2	74	135	275	[25]
Mg_2FeH_6	80	137	311	[26]
$LiBH_4$	74	115	370	[27]
(1)$NaAlH_4$	42	124	66	[28]
$LiNH_2$	45	n.a.	-	[29]
(2)$2LiBH_4$+MgH_2	40.5	81.3	255	[30]
$Mg(NH_2)_2$+2LiH	38.9	107(3)	90	[31]

(1) The values are reported as an average of the two steps reaction from NaH and Al to $NaAlH_4$ [28].
(2) Values obtained from hydrogenation PCIs, assuming one reaction step [30].
(3) Entropy calculated from the reported van't Hoff plot [31]. n.a. = not available.

Table 2.
Dehydrogenation thermodynamic parameters and decomposition temperature under 1 bar H_2 for some hydride compounds and systems.

H_2 for some hydride compounds and systems, which have been under intensive research [24–31].

Table 2 clearly indicates that the use of some hydride compounds and systems for solid-state hydrogen storage system working under mild temperature conditions is thermodynamically feasible. Moreover, other cases such as Mg_2FeH_6 are more suitable for energy storage because of its large enthalpy value. All efforts toward the design of hydride compounds, especially for mobile applications, have been focused on reaching enthalpy values ranging between 40 and 50 kJ/mol H_2, leading to an operative temperature between 20 and 90°C. It seems that some materials such as room temperature hydride fulfill these requirements, but the thermodynamic stability is not the unique parameter that determines the practical application of the hydride compound, since the capacity plays a major role. For this reason, high gravimetric hydrogen capacities are required as well, and room temperature hydrides do not meet this requirement at the time to work at low hydrogen pressure.

3. Fundamental concepts about hydrogenation-dehydrogenation kinetics

Most of the hydride compounds and hydride systems cannot work at the temperature predicted by the thermodynamics. This fact is related to kinetic constraints, leading to notable higher operative temperatures than that thermodynamically feasible. For this reason, enhancing the kinetic behavior of hydride compounds and hydride systems has been matter of intensive research. To start with a comprehensive development about the strategies to improve the kinetic behavior of the hydride compounds and systems, it is important to introduce fundamental concepts and details about the experimental method used to understand and to tailor the hydrogenation/dehydrogenation rates, respectively.

The kinetic behavior for the formation/decomposition of a hydride compound gives information about the time required for the material to uptake/release hydrogen in non-equilibrium (dynamic) conditions. **Figure 4** shows experimental kinetic hydrogenation and dehydrogenation curves for Nb_2O_5-doped MgH_2. These curves show hydrogen concentration as a function of time. For the sake of clarity, all the explanation and description is referred to a nominal metal hydride ($MH_{x(s)}$, M = metal, H = hydrogen). The absorbed/desorbed hydrogen concentration is expressed in terms of fraction (α), which is the ratio between the absorbed/desorbed hydrogen concentration at each time during the process and the

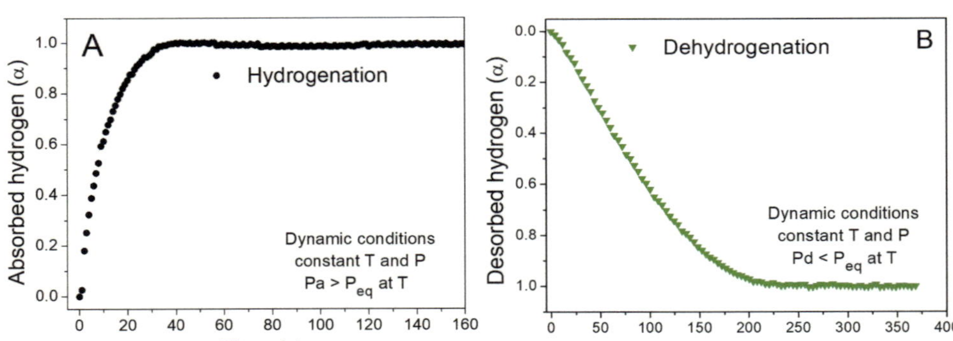

Figure 4.
(A) Hydrogenation kinetic behavior and (B) dehydrogenation kinetic behavior for metal-hydrogen system.

maximum absorbed/desorbed hydrogen concentration. Under dynamic conditions, the hydrogenation reaction happens when the operative pressure (Pa) is over the equilibrium pressure (P_{eq}) of the hydride phase at the operative temperature (T), **Figure 4A**. On the contrary, for dehydrogenation under dynamic conditions, the operative pressure (Pd) has to be below the equilibrium pressure (P_{eq}) of the hydride phase at the operative temperature (T), **Figure 4B**. The procedure to measure a kinetic curve for the hydrogenation/dehydrogenation consist in introducing the hydride forming metal or material in a sealed vessel connected to hydrogen supply and suddenly increase/decrease the hydrogen pressure at a constant temperature, respectively. After the sudden pressure increase/decrease, the operative pressure is kept constant and higher/lower than the P_{eq} for the hydrogenation/dehydrogenation dynamic processes, respectively. The measurement of the kinetic curve runs until reaching saturation, i.e., $\alpha = 1$.

In order to clarify the complexity of the hydrogenation/dehydrogenation processes under dynamic conditions, the sequence of involving steps can be simply described as follows:

3.1 Hydrogenation process

a. **Physisorption:** The hydrogen molecule approaches to the metal and interacts with the metal's surface through van der Waals forces.

b. **Chemisorption:** On the metal's surface, the hydrogen molecules dissociate and the atomic hydrogen forms chemical bonds with the metal.

c. **Penetration through the α phase surface:** Hydrogen atoms penetrate the metal's surface.

d. **Hydrogenation:** Formation of the β phase (hydride phase).

e. **Diffusion through the β phase:** Hydrogen atoms diffuse through the β phase (hydride phase).

f. **Nucleation and growth:** The β phase (hydride phase) starts to nucleate and grows with α/β interface movement.

The **dehydrogenation process** proceeds in the opposite way, as described above. In **Figure 5**, the involved steps during the hydrogenation and dehydrogenation processes in dynamic conditions are shown.

In the previous description (**Figure 5**), any fluid dynamic and heat transfer restrictions associated to large amounts of material are not taken into account. For instance, fluid dynamic restriction occurs when hydrogen can flow into the material's bed. For the case of the heat transfer restrictions, it is referred to the notable temperature increase or decrease during the exothermic or endothermic hydrogenation or dehydrogenation reaction, respectively. The heat transfer through the hydride bed is not efficient because the thermal conductivity of the hydride compounds is rather low. For these reasons, the analysis on the kinetic behavior and tailoring will be performed for a punctual mass of hydride bed without any fluid dynamic and heat transfer restriction.

There are functional dependences of the hydrogenation/dehydrogenation reaction rates on the temperature, pressure, and morphology of the material. The rates for gas-solid reactions, such as the ones occurring during the hydride formation and decomposition, can be described by Eq. (5) [32, 33]:

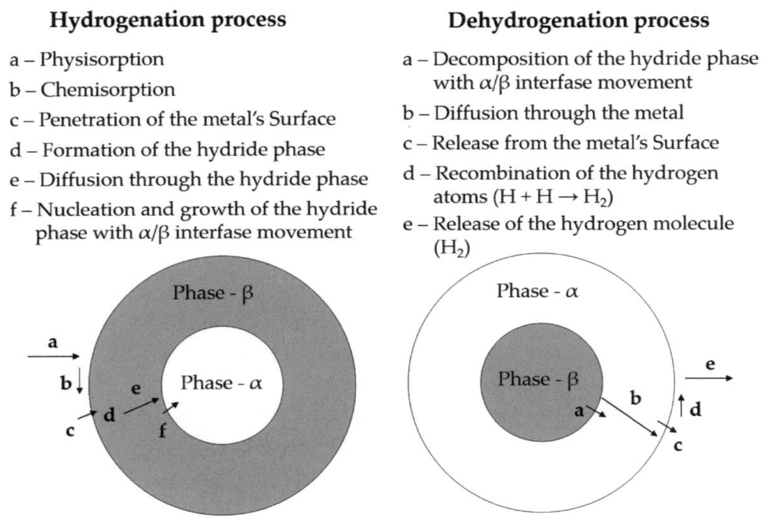

Figure 5.
Steps involved in the hydrogenation and dehydrogenation processes under dynamic conditions.

$$\frac{d\alpha}{dt} = K(T) \times F(P) \times G(\alpha), \tag{5}$$

where the overall reaction rate, $d\alpha/dt$ (α = hydrogen fraction and t = time) is function of the temperature, K(T), of the pressure, F(P), and of the intrinsic and morphological changes of the material occurring during the hydrogenation/dehydrogenation process, $G(\alpha)$, which is function of the hydrogen fraction (α). As shown by Eq. (5), the dependencies of the rate on the K(T), F(P), and $G(\alpha)$ can be investigated independently by keeping two of them constant. In the following sections, each dependence will be explained.

3.2 Dependence on the intrinsic and morphological changes of the material G(α)

At constant temperature and pressure, the reaction rate depends on intrinsic factors and morphological changes of the solid (defects, crystalline structure, size and morphology of the particles, etc.), i.e., $G(\alpha)$ [34, 35]. Thus, Eq. (5) can be expressed:

$$\frac{d\alpha}{dt} = k \times G(\alpha) \tag{6}$$

$$k = K(T) \times F(P), \tag{7}$$

k is the kinetic constant and it contains the temperature and pressure dependences as seen in Eq. (7). Reordering Eq. (6), it is possible to obtain an expression for the evaluation of $G(\alpha)$:

$$g(\alpha) = \int_0^\alpha \frac{d\alpha}{G(\alpha)} = k \times t, \tag{8}$$

where $g(\alpha)$ is the integral form of the gas-solid kinetic models for the material's changes. The gas-solid kinetic models describe different physical phenomena. There are four main sets of models: (1) nucleation and growth models, (2) geometrical contracting models, (3) diffusion models, and (4) autocatalytic models. In **Table 3**, the integral form of the gas-solid kinetic models is described. There are more gas-solid models, but it is not the aim to perform exhaustive description of them.

Among the steps involved in the hydrogenation/dehydrogenation processes (**Figure 5**), it is usually found a slowest one, which limits the overall reaction rate of the process. Thus, the slowest step is commonly called "rate-limiting step." The determination of the rate-limiting step depends on the kind of hydride and the experimental conditions. Identifying the rate-limiting step requires the application of gas-solid models shown in **Table 3**. The determination of the rate-limiting step is carried out by measuring kinetic curves under constant temperature and pressure, as the ones shown in **Figure 4**. Once, the hydrogen uptake and release against time is expressed in terms of hydrogen fraction, α, the integral models can be applied to build a graph of g(α) as function of time.

A first approach to study which process limits the hydrogenation/dehydrogenation kinetic behavior is to find which model has the best linear fitting of the integral

Model	Description	Integral form $g(\alpha) = k \times t$
Nucleation and growth models		
Johnson-Mehl-Avrami (JAM), n= 1	One-dimensional growth with interface-controlled reaction rate	$[-\ln(1-\alpha)]^{1/n}$
Johnson-Mehl-Avrami (JAM), n= 2	Two-dimensional growth of existent nuclei at constant interface rate	
Johnson-Mehl-Avrami (JAM), n= 3	Three-dimensional growth of existent nuclei at constant interface rate	
Johnson-Mehl-Avrami (JAM), n= 1.5	Three-dimensional growth of random nuclei with decreasing interface rate, diffusion controlled	
Geometrical contracting models		
Contracting area (CA), n = 2	Two-dimensional growth of contracting volume with constant interface rate	$1-(1-\alpha)^{1/n}$
Contracting volume (CV), n = 3	Three-dimensional growth of contracting volume with constant interface rate	
Diffusion models		
1-D Diffusion	Surface control (Chemisorption)	α^2
2-D Diffusion	Two-dimensional diffusion controlled growth with decreasing interface rate	$[(1-\alpha)\ln(1-\alpha)]+\alpha$
3-D Diffusion	Three-dimensional diffusion controlled growth with decreasing interface rate	$1-\left(\frac{2}{3}\right)\alpha-(1-\alpha)^{2/3}$
Autocatalytic models		
Modified Prout-Tompkins	Acceleration of the reaction rate with product (s) generation. a = constant	$ln\left[\frac{a}{(1-a)}\right]$

Table 3.
Integral form of the gas-solid models [32, 33, 36].

form, g(α), of the solid-state models against time. First, the hydrogen fraction range considered for the fitting is usually from 0.1 to 0.9. On one hand, at the beginning of the process the initial points involve a high degree of uncertainty, mainly for fast rates in the range of seconds. On the other hand, the final stage of the process, reaching the saturation of the material, does not play any role in the determination of the rate-limiting step. As an example, **Figure 6** exhibits g(α) generated from each model, **Table 3**, as a function of time in the range of α between 0.10 and 0.90 for the hydrogenation and dehydrogenation process shown in **Figure 4**. Then, a liner fitting was performed in each curve.

As seen, the simple linear fitting of the integral form of the solid state-models does not provide a clear clue about the rate-limiting step for the hydrogenation and dehydrogenation rates. There are several models with suitable fitting goodness (highlighted in bold). For these reason, it is possible to apply other procedure such as the reduced time method proposed by Sharp and Jones [37, 38]. By applying this method, a plot for the theoretical reduced time (t/t0.5)theoretical as a function of the experimental reduced time (t/t0.5)experimental is built as shown in **Figure 7**. The (t/t0.5)theoretical is obtained by expressing the equation of the integral form of solid-state models in terms of the time at α = 0.5, i.e., t0.5. As an example, for the case of the integral form of the solid-gas model JMA with n = 1, the (t/t0.5)theoretical is obtained as follows:

$$[-\ln(1-\alpha)] = k \times t \tag{9}$$

$$[-\ln(1-0.5)] = k \times t_{0.5} \tag{10}$$

The right term of Eq. (10) is equal to 0.69. Replacing 0.69 in Eq. (10) and dividing (9) over (10), Eq. (11) provides the $(t/t0.5)_{\text{theoretical}}$

$$\frac{[-\ln(1-\alpha)]}{0.69} = \left[\frac{t}{t_{0.5}}\right]_{theoretical} \tag{11}$$

The (t/t0.5)experimental is directly obtained from the experimental results, **Figure 4**, by dividing the time during the measurement (t) over the time at α = 0.5. The rate-limiting step is determined by three parameters, i.e., the fitting goodness, slope, and intercept of the linear fitting. This represents an advantage in comparison with the simple linear fitting of the integral form of the solid-state models, with just the fitting goodness as one decision parameter. The best fitting obtained from

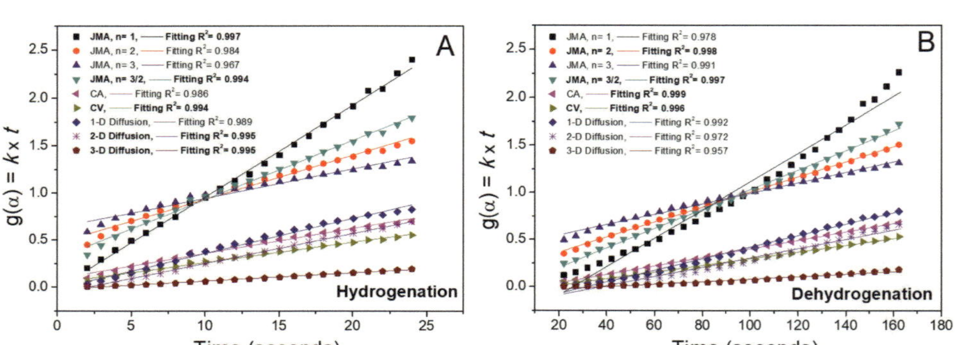

Figure 6.
*Application and fitting of the integral form of the solid-state models (**Table 3**) to the hydrogenation/dehydrogenation rates shown in **Figure 4**.*

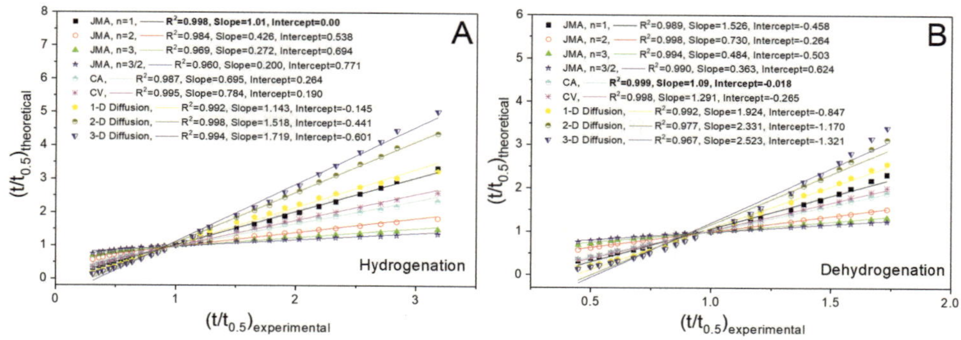

Figure 7.
Reduced time method: Plots $(t/t_{0.5})_{theoretical}$ vs. $(t/t_{0.5})_{experimental}$ and fitting for (A) hydrogenation and (B) dehydrogenation rates.

the reduced time method is the one showing linear fitting goodness and slope equal to 1 and intercept equal to 0.

Figure 7 shows that for the case of the hydrogenation rate the model JMA, n = 1, **Table 3**, provides the best linear fitting (slope = 1, intercept = 0 and goodness ≈ 1). For this reason, the overall reaction rate is limited by one-dimensional growth with interface-controlled reaction rate. This rate-limiting step is related to the step (f) described in **Figure 5**. For the dehydrogenation rate, the rate-limiting step is the two-dimensional growth of contracting volume with constant interface rate (CA), related to the step (a) shown in **Figure 5**.

In fact, the kinetic analysis based on the solid-state models should be also based on additional experimental evidence obtained from the material science characterizations for morphological and microstructural changes before and after and upon hydrogenation and dehydrogenation kinetic processes.

3.3 Temperature dependence $K(\mathrm{T})$

Under constant pressure, the hydrogenation/dehydrogenation reactions speed up as the temperature increases and this temperature dependence is described by the Arrhenius expression. The dependence of the hydrogenation/dehydrogenation reaction rates on the temperature follows Eq. (12):

$$K(T) = A \times \exp\left(-{}^{Ea}\!/_{RT}\right), \tag{12}$$

where A is the pre-exponential factor or frequency factor and represents the frequency of collisions between reactant molecules, Ea is the apparent activation energy, which is the energy required to start the reaction, R is the gas constant, and T is the absolute temperature.

The activation energy is a relevant kinetic parameter to evaluate the kinetic performance of a hydride forming material. Isothermal and non-isothermal measurements can be performed to obtain the activation energy. Two procedures are commonly used. The first consists in performing isothermal measurements at different temperature and constant pressure, such as the ones shown in **Figure 4**. Then, by the application of the gas-solid models, it is possible to calculate the kinetic constant "k," once identified the rate-limiting step as explained in Section 3.1. As expressed in Eq. (7), k depends on K(T) and F(P). Considering isobaric measurements, k is only function of K(T) following the Arrhenius expression:

$$k = A \times \exp\left(-{}^{Ea}\!/_{RT}\right), \tag{13}$$

Applying natural logarithm to Eq. (13):

$$lnk = lnA + ln\left(^{-Ea}/_{RT}\right) \quad (14)$$

Plotting lnk against 1/T, the slope of the linear fitting provides the Ea in kJ/mol H2 and the intercept the frequency factor A, as seen in **Figure 8A**. The second method involves non-isothermal measurements done via calorimetric methods such as scanning differential calorimetry (DSC) and differential temperature analysis (DTA) [39]. This method is the Kissinger one [40]. The method consists in determining temperature maxima (Tm) of the exothermal (hydrogenation)/ endothermal (dehydrogenation) events obtained from the DSC or DTA measurements at different constant heating rate (ø). The change of Tm with ø is directly related to the nature of the reaction.

At a constant heating rate (ø), constant pressure (F(P)) and considering k has an Arrhenius dependence on the temperature, Eq. (13), Eq. (5) can be expressed:

$$\frac{d\alpha}{dt} = \frac{A}{dt} \times \exp\left(^{-Ea}/_{RT}\right) \times G(\alpha) \quad (15)$$

Assuming that the fraction at the peak maximum and G(α) are independent on ø, reordering and applying natural logarithm to expression (15), it is possible to write:

$$ln\left[\frac{\varnothing}{T_m^2}\right] = -\frac{Ea}{RT_m} + ln\left[\frac{AR}{Ea}\right] + C \quad (16)$$

Figure 8B shows the DSC curves at different ø for a dehydrogenation process, from which the peak maxima are selected. Then, a plot of the ln(ø/Tm2) against 1/T is built and the Ea in kJ/mol H_2 is obtained from the slope of the linear fitting, as shown in **Figure 8C**.

Table 4 shows examples for experimental activation energy (Ea) values, reaction mechanism, and rate-limiting steps for different materials. As seen, the lowest Ea values are related to surface controlled process such as physisorption and chemisorption (**Figure** 5: Hydrogenation process—Steps **a** and **b** and Dehydrogenation process—Steps **d** and **e**). On the contrary, bulk processes such as interface controlled and diffusion processes present higher *Ea* values (**Figure** 5: Hydrogenation process—Steps **c**, **e**, and **f** and Dehydrogenation process—Steps **a** and **b**). The

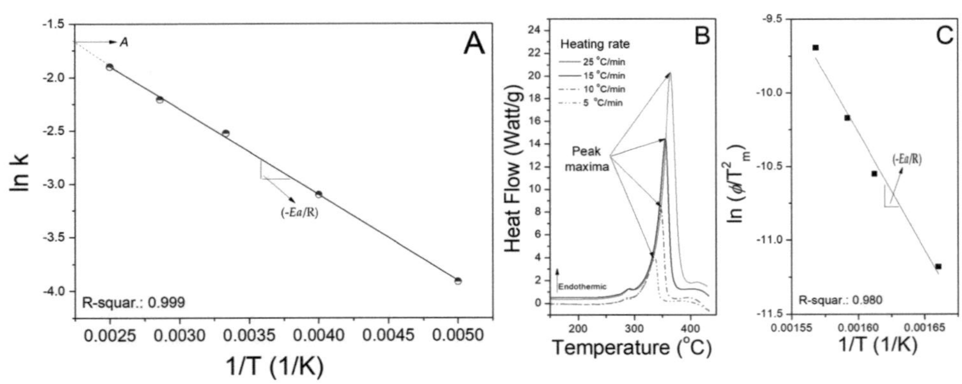

Figure 8.
Activation energy determination: (A) isothermal method, (B) DSC curves for the (C) non-isothermal method.

Ea [kJ/mol H_2]	Process, material, physical meaning,rate-limiting step	Ref.
54.1	Dehydrogenation for $LaNi_{4.7}Al_{0.3}$–H. Rate limiting step: Nucleation and Growth (α+β phases).	[41]
28.9	Dehydrogenation for $LaNi_5$. Rate-limiting step: Surface control and chemical reaction.	[42]
40±2	Dehydrogenation $LaNi_5H_x$. Shrinking core model. Rate-limiting step: Controlled interface reaction at the β/α interface.	[43]
9.1	Hydrogenation for $LaNi_5$. Rate-limiting step: Surface control. H_2 molecule dissociation on the surface of the β phase (high temperature)	[44]
45.5	Hydrogenation for $LaNi_5$. Rate-limiting step: Bulk diffusion control. Diffusion of hydrogen atoms through the β phase (low temperature)	
19	Hydrogenation for $Ti_{0.98}Zr_{0.02}V_{0.43}Fe_{0.09}Cr_{0.05}$. Rate limiting step: Diffusion of hydrogen atoms.	[45]
62	Dehydrogenation for Mg_2Ni. Rate-limiting step: Diffusion of hydrogen through the α phase.	[46]
12.8	Dehydrogenation for $MnNi_{4.5}Al_{0.5}$. Rate-limiting step: Chemisorption on the active sites and α/β interface.	[47]
46.5	Hydrogenation for $MnNi_{4.5}Al_{0.5}$. Rate-limiting step: Hydrogen diffusion through the hydride (β) phase.	
39.9	Hydrogenation for $LaNi_5$. Rate-limiting step: Hydrogen diffusion through the hydride (β) phase.	[48]
33.0	Hydrogenation for $LaNi_5$. Rate-limiting step: Controlled interface reaction at the β/α interface.	[49]
0.92	Hydrogenation for Mg_2Ni. Dissociative chemisorption of the H_2 molecule. Segregated and excess nickel as active sites for dissociation.	[50]
90±10	Hydrogenation for Mg/MgH_2. Rate-limiting step: Nucleation and growth mechanism for the β phase.	[51]
100±10	Dehydrogenation for MgH_2. Rate-limiting step: hydrogen diffusion through the β phase. Corrected by *F*(P)	
160±10	Hydrogenation of $2LiH+MgB_2+0.05molTiCl_3$. Rate limiting step: One-dimensional interface controlled. Corrected by *F*(P).	[52]
350±10/238±4	Dehydrogenation for $2LiBH_4+MgH_2+0.05molTiCl_3$. First step: dehydrogenation of MgH_2/Second step: Dehydrogenation of $LiBH_4$	

Table 4.
Activation energies, reaction mechanisms, and rate-limiting steps.

interface and diffusion controlled processes are the most common ones for the hydride compounds formation/decomposition [41–51].

It is important to point out that the isothermal method for *Ea* requires the knowledge of the rate-limiting step, thus it depends on the solid-state model. On the contrary, the non-isothermal method is independent on the solid-state model, which represents an advantage over the first method. However, for the kinetic modeling of the reaction rates, as indicated by Eq. (5), the determination of G(α) is demanding. For this reason, at the time proceed with kinetic modeling, or just determine *Ea via* measurements of the hydrogenation or dehydrogenation rates at isothermal conditions, it is necessary to work far away from the equilibrium pressures to avoid the effects of the driving force, i.e., the *F*(P). In fact, rightly speaking, the *K*(T) must be corrected by the effects of the driving force by determining *F*(P), since the *Ea* values obtained after the correction are different from the one without; as an example we can mention the works published by Fenández et al. [51] and Jepsen et al. [52]. In the next subsection, then, the F(P) dependency is described.

3.4 Pressure dependence F(P)

At constant temperature, the hydrogenation/dehydrogenation reactions show a dependence on a functional relationship between the operative pressure (P) and the equilibrium pressure (Peq) for both hydrogenation and dehydrogenation processes. This dependence is called *F*(P). **Table 5** describes the most relevant *F*(P) applied to different materials and their physical meaning. These functions were obtained from experimental investigations and proposed in pioneering works about the kinetic

F(P)	Physical meaning / rate-limiting step / material	Ref.
$(P/Peq)^n$, n=1	*F*(P) developed considering the reversible nature of the hydrogenation/dehydrogenation reactions in $LaNi_{4.7}Al_{0.3}$ alloy. Rate-limiting step: Nucleation and growth in the $\alpha+\beta$ phase.	[41]
$(P - Peq)/P_o$	Chemical reaction as rate-limiting step. It was determined for the dehydrogenation reaction of $LaNi_5H_x$ as the best-fitted function. P_o= 1 bar (reference pressure).	[42]
ln(P/Peq)	$(\beta)LaNi_5H_6 \rightarrow (\alpha)LaNi_5H_x+(3-x/2)H_2$: Isothermal dehydrogenation. Shrinking core model, with interphase controlled as rate-limiting step, assuming that the surface is perfectly porous, so that neither diffusion nor mass transport can be rate-limiting steps. Therefore, the chemical potential acts across the β/α interface.	[43]
	Hydrogenation of $LaNi_5$: Diffusion of H atoms from the surface through the growing β phase to the α/β interface.	[44]
(Peq − P)	Dehydrogenation of $LaNi_5$: Surface controlled as ratel-limiting step (Recombination of the H_2 molecule on the α surface.	
$P^{0.5}$	Hydrogenation for $Ti_{0.98}Zr_{0.02}V_{0.43}Fe_{0.09}Cr_{0.05}$. Diffusion of H–Atom as rate-limiting step.	[45]
$(Peq^{0.5} - P^{0.5})$	Dehydrogenation for the Mg_2Ni alloy (Mg_2NiH_4). Rate-limiting step: Diffusion of hydrogen through the α phase at low and high pressures.	[46]
(P − Peq)	Hydrogenation for $MnNi_{4.5}Al_{0.5}$. Rate-limiting step: Chemisorption on the active sites and α/β interface.	[47]
$(P^{0.5} - Peq^{0.5})$	Hydrogenation for $MnNi_{4.5}Al_{0.5}$. Rate-limiting step: Hydrogen diffusion through the hydride phase.	
(P − Peq)	Hydrogenation for Mg_2Ni. Rate-limiting step: Dissociative chemisorption of the H2 molecule.	[50]
$(1 - P/Peq)^2$	Hydrogenation for Mg/MgH_2. Nucleation and growth process. Rate-limiting step: JMA with n between 0.5 to 1	[51]
P	At room temperature, the pressure dependence of the H_2 reaction on Ti surfaces contaminated with CH_4, CO, CO_2, H_2O, O_2, N_2 and SO_2 is approximately proportional to P (operative pressure).	[53]
$(Peq - P_v)/Peq$	$TiFe_{0.8}Ni_{0.2}$: Investigations on kinetics after prolonged cycling. No physical meaning. P_v: instantaneous differential pressure in the collecting desorption reservoir.	[54]
$1 - (P/Peq)^{0.5}$	Hydrogenation for Mg/MgH_2. Rate-limiting step: Diffusion	[55]

Table 5.
Pressure dependencies, their physical mining, and rate-limiting step.

behavior of hydride compounds [41–47, 50, 51, 53–55]. Moreover, the shown *F*(P) were also applied to characterize the kinetic behavior and to model the reaction rates of different hydride compounds and hydride systems; exemplary for $NaAlH_4$ complex hydride with Ti-based catalyst [56, 57]. As seen, the developed *F*(P) depends on the hydride forming material and each *F*(P) is associated to different rate-limiting step and physical meaning. It is important to mention that the experimental conditions also play a major role for the pressure dependence, thus it is possible to find works in which the same material was investigated, but the pressure dependences are different, and this fact is mainly related to different experimental setups.

In this section, all the basics concepts for the right understanding of the kinetic behavior of the hydride compounds and dependencies of the hydrogenation/dehydrogenation reaction rates were exposed. Now, it is possible to describe the main strategies applied to improve the hydrogenation/dehydrogenation rates for some hydride compounds and hydride systems under extensive research.

4. Strategies to tailor the kinetic behavior of hydride compounds and systems

Investigations on the kinetic tailoring to improve the hydrogen storage properties of hydride compounds and hydride systems involve knowledge about the thermodynamic behavior and kinetic behavior of the hydride system. First, equilibrium pressures at different temperatures are needed to plan the experimental pressure and temperature for the characterization of the kinetic behavior. Second, some knowledge about the mechanism(s) that constraint the overall reaction rates is also required. For this reason, it is important to know or at least have clues about possible rate-limiting steps for the hydrogenation/dehydrogenation processes, as well as the influence of the temperature and pressure dependences on the reaction rates. In this regard, knowledge about K(T) allows quantifying in terms of Ea the effects of a tailoring strategy. In this line, knowledge of F(P) dependence is usually necessary to properly determine K(T). All these understanding about the thermodynamics and kinetics of the hydride compound or system matter of investigation enormously contribute to seek the best strategy to improve it hydrogenation/dehydrogenation reaction rates. Of course, there are several opportunities in which kinetic improvements for a hydride compound or system are proposed based just on a simple experimental procedure; without any previous systematic study about the thermodynamics and kinetics. This approach is also possible, but it leads to a superficial knowledge about the actual kinetic improvement and to a poor understanding of the behavior of the hydride compound or system.

The most relevant approaches to enhance the kinetic behavior of hydride compounds and systems are: (1) improving the microstructural refinement via mechanical milling, (2) doping with transition metal and transition metal compounds, (3) forming in situ catalyst, and (4) nanoconfinement. In this section, the concepts of these strategies are explained and examples for their practical application to hydride compounds and systems are also described [58–96].

In general, two or three of these strategies are applied together, as for example mechanical milling, transition metal, or transition metal compound addition and in situ catalyst formation or nanoconfinement and transition metal or compound addition, etc. However, it is possible to put emphasis on the each strategy in order to understand what the specific effects are on the hydride forming, regardless the combined application of them.

4.1 Mechanical milling

The mechanical milling process is a powder processing method. This processing technique allows producing homogeneous materials starting from powder mixtures [58]. Moreover, the mechanical milling causes particle and grain size reduction (microstructural refinement). Furthermore, mechanical milling is also applied for the synthesis of hydrides, i.e., mechanochemical synthesis [59]. This process can be carried out at laboratory scale and at industrial scale. Herein, the laboratory scale milling process of hydride compounds and systems are discussed. In general, the milling process for improving the microstructural characteristics of hydrides is performed with amounts of material ranging between 1 and 20 g. The milling vessels are usually made of stainless steel and with a volume between 50 and 250 cm^3. As grinding medium stainless steel balls are used. There are several laboratory milling devices that can be classified by the injected shock power per unit mass of grinding medium (Pg), which is a parameter that determines the reached microstructural refinement of a material (powder) subjected to milling procedure. Indeed, the reached degree of microstructural refinements also depends on the milling time and the numbers of ball (grinding medium), which are parameters that can be changed at the time to perform the milling process. Among the most commonly utilized mill devices for hydride compounds and system preparation at laboratory scale, it is possible to mention: (1) Magneto Uni-ball mill device (low energy, Pg: 0.0003–0.002 W/g), (2) Planetary Fritsch-P6 (middle energy, Pg: 0.01–0.22 W/g), and Vibratory Spex 8000 M mill device (high energy, Pg $<$ 0.24 W/g) [60, 61]. Owing to the pyrophoric nature of the hydride compounds and their easiness to get hydrolyzed and/or oxidized, the milling process is done under inert atmosphere or hydrogen atmosphere in the case of mechanochemical synthesis. The effects of mechanical milling on the material are: (1) creation of grain boundaries and defects, (2) higher degree of microstructural refinement, and (3) intimate mixture of powders. In this regard, the first two effects improve the activation process of the material for the initial hydrogen absorption and the hydrogenation/dehydrogenation kinetic processes due to the enhancement of hydrogen diffusion through new pathways and shorter distances. Grain boundaries and defects are created by the mechanical energy provided during the milling process. More grain boundaries provide larger surface area and shorter diffusion distance for hydrogen, while the presence of defects acts as channels for hydrogen diffusion. Both effects ease the contact between hydrogen and fresh hydride forming material, enhancing the kinetic behavior. The third effect contributes to the homogeneous distribution and intermixture of the main components of the material and an additive. Moreover, the milling process can promote interactions between the main components and the additive, leading to the formation of other species with favorable effects on the kinetic behavior. This effect will be addressed in a following section. **Figure 9** shows the effects of mechanical milling under hydrogen (0.5 bar) atmosphere for 150 h on a mixture of Mg + 10 wt.%Fe in a Magneto Uni-ball mill device. As seen in the SEM photos (Scanning Electron Microscopy), the particle sizes are notably reduced and the surface has porous characteristic in comparison to the starting material. These results account for the mechanical effects of the milling process as well as the fragilization of Mg (ductile material) because of the *in situ* formation of MgH_2 (brittle material) under H_2 atmosphere. The effects of milling on Mg + 10 wt.%Fe markedly improve its kinetic behavior by reducing the diffusion pathways for the hydrogen absorption [62].

In 1997, Zaluska et al. [63] reported a pioneering work about the properties of different as-milled hydride forming materials such as Mg, Mg_2Ni, FeTi, and $LaNi_5$. It was found that the materials after milling presented a nanocrystalline nature,

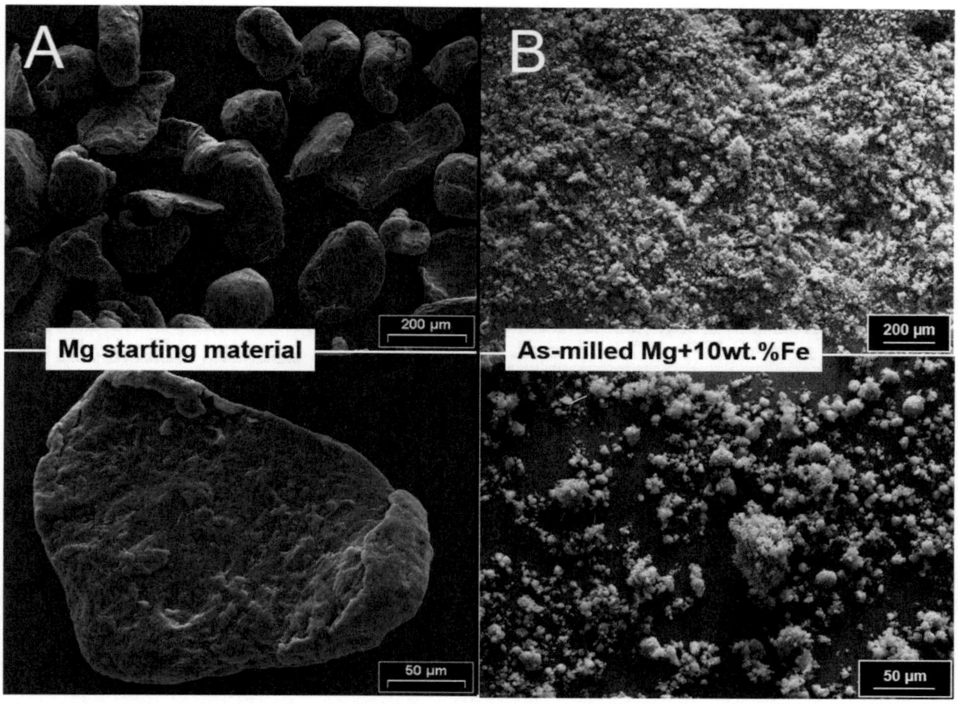

Figure 9.
SEM photos for (A) Mg starting material and (B) Mg + 10 wt.%Fe after 150 h of milling under 0.5 bar H_2 in a magneto Uni-ball device.

causing notable improvements in the hydrogenation/dehydrogenation kinetic behavior. These effects are also noticed in complex hydrides and hydride system [64]. Despite the beneficial effects of milling, there are also some disadvantages. In the case of room temperature hydrides, the effects of milling are not always beneficial. For example, it was also reported that FeTi alloys undergo hydrogen capacity loss after milling process. This is due to the creation of amorphous regions induced by the mechanical deformation and it is not possible to store hydrogen interstitially in the amorphous regions [65]. Furthermore, taking into account that the milling vessel and grinding medium are usually made of stainless steel, Fe contamination is also a concern, particularly at the time to perform the milling process in a high-energy mill device. For instance, as demonstrated by Puszkiel et al., the presence of Fe for the $2LiBH_4$ + MgH_2 hydride system causes detrimental effects in the kinetic behavior since FeB species without any beneficial effects are formed. It was found that an appreciable amount of Fe came mostly from the grinding medium [66]. Despite the described disadvantages, the milling process is quite effective and efficient at the time to prepare and tailor hydride compounds and systems and additionally it is possible to scale-up for practical applications. Reducing the grain and particle sizes, creating more boundary surfaces and defects are the most relevant effects of the milling process. These effects mainly improve the diffusion processes for the hydride formation and decomposition by shortening the diffusion pathways and generating channels for more efficient hydrogen transport.

4.2 Doping with transition metal and transition metal compounds

Since the beginning of the investigations on hydride compounds, it was discovered that the addition of transition metal and metal compounds were able to improve their kinetic characteristics. The addition of amounts from 1 to 10 wt.% of these transition metals and compounds was performed by milling procedures in most of the cases. Therefore, the "doping strategy" also involves the milling process to reach a high degree of intermixture and microstructural refinement. The doping strategy results in reductions of the hydrogenation and/or dehydrogenation activation energies, changes in the reaction paths and consequently different rate-limiting steps. **Figure 10** shows the concept of doping *via* mechanical milling. As an example, the activation energy for the dehydrogenation process for a bulk hydride MHx (M = metal) is reduced after adding a transition metal and transition metal compound by milling process, $Ea_1 > Ea_2$.

In the case of room temperature hydrides, the presence of transition metals like Pd and Ni notably improves their activation behavior and kinetic properties. This fact was attributed to the active sites of these transition metals, located on the metal or alloy surface, which facilitated the hydrogen molecule dissociation and penetration across the oxides generated on the metal or alloy surface [67–69]. These hydrides are commonly prepared by arc melting since, as mentioned in Section 4.2, the milling process can cause hydrogen capacity losses.

In 1997, Bogdanović et al. [70] achieved reversible hydrogen uptake and release from $NaAlH_4$ under mild conditions by doping with Ti-based catalyst ($TiCl_3$). However, the catalytic mechanism of the Ti-based catalyst was not clear at those times. In 2015, Züttel et al. [71] reported a work about the catalytic mechanism of Ti-compound in the hydrogen uptake/release of alkali alanates. In this work, based on an atomistic model, it was proposed that Ti works as a bridge to transfer H^- and M^+ (M^+---Ti---H^-) from $MAlH_4$ (M = Li, Na, K), reducing their charge separation and thus lowering the activation energies for hydrogenation and dehydrogenation processes.

Among the most interesting hydrides, MgH_2 is very attractive because of its low cost and high gravimetric hydrogen capacity (7.6 wt.%). However, its high thermodynamic stability (74 kJ/mol H_2) required high operative temperature over 300°C [25]. Pure MgH_2 has sluggish kinetic behavior. Upon hydrogenation, the rate-limiting step is three-dimensional diffusion controlled contracting volume. This fact is attributed to the low diffusion coefficient of MgH_2 covering fresh Mg; the diffusion coefficient of MgH_2 was found to be three orders of magnitude lower that the one for Mg). Upon dehydrogenation, the rate-limiting step for pristine

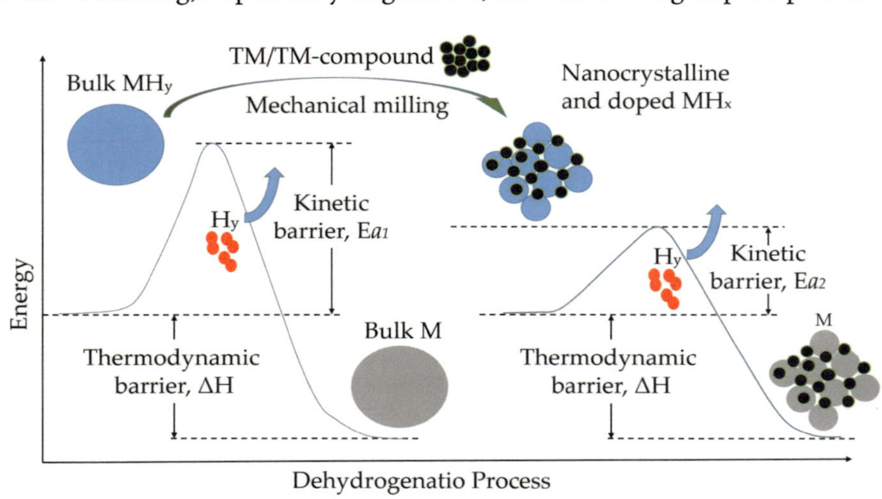

Figure 10.
Concept of doping via mechanical milling. TM = transition metal. MHy = hydride, M = metal, ΔH = enthalpy, Ea = *activation energy.*

MgH_2 turned to be surface controlled and related to the recombination of the hydrogen molecule [72, 73]. Several transition metals and transition metal compounds exhibited excellent catalytic effect on de-/hydrogenation of MgH_2. The main effect of these additives was related to the dehydrogenation mechanism, since they ease the recombination of the hydrogen molecule [74, 75]. One of the most effective additives was Nb_2O_5, leading to 7 wt.% H_2 capacity in about 60 s and fast dehydrogenation in about 130 s. Moreover, the presence of Nb_2O_5 reduced the dehydrogenation temperature down to 250°C [76]. Nb_2O_5 had a "pathway effect" mainly on the hydrogenation of MgH_2. Upon milling and subsequent hydrogen cycling, the formation of Mg-Nb oxides created diffusion pathways by the formation of metastable niobium hydrides, hence avoiding the large diffusion constraints. The effect and mechanism of Nb_2O_5 on MgH_2 is similar but slower when MgH_2 is physically doped with Nb_2O_5 [77, 78].

$LiBH_4$ also caught the attention due to the extremely high gravimetric capacity of 18.3 wt.% H_2, though just 13.8 wt.% H_2 is available because LiH and B are formed upon desorption. However, $LiBH_4$ is quite stable and needs harsh temperature and pressure conditions for de-/re-hydrogenation [79]. Several dopants such as metal oxides and metal halides were tried for improving the kinetic behavior of $LiBH_4$. However, $LiBH_4$ reacts with the dopants because this complex hydride is a strong reduction agent. The interactions between $LiBH_4$ and dopants also lead to gas byproducts such as B_2H_6. In this regard, complex hydrides such as borohydrides and amides are combined with binary hydrides to form "thermodynamically destabilized systems" and then the kinetic behavior of these hydride systems is tailored by doping [30, 31]. In almost all the cases, the added dopants interact with the hydride system itself by forming other species *in situ*, and this strategy will be addressed in the next section.

4.3 *In situ* catalyst formation

This strategy is applied to binary hydrides, to complex hydrides and mainly to destabilized hydride systems. In this section, the *in situ* formation of species with catalytic activity for destabilized hydride system is described. These hydride systems present proper thermodynamic parameters as for example $2LiBH_4 + MgH_2$ and $Mg(NH_2)_2 + 2LiH$, as shown in **Table 2**, because of the exothermic formation of reversible species during the endothermic dehydrogenation. Nonetheless, these hydride systems still present kinetic constraints to reach the operative conditions predicted by the thermodynamic. One of the main strategies to overcome this problem is the addition of dopants. In almost all the cases, these dopants interact with the hydride system by forming *in situ* species with catalytic activity.

One of the most representative examples of this approach is the addition of transition metal compounds to the $2LiBH_4 + MgH_2$ hydride system. Under dynamic conditions, this borohydride system uptake hydrogen in one step, but releases hydrogen in two steps as described in reactions (17)–(19), respectively [80].

$$2LiH_{(s)} + MgB_{2(s)} + 4H_{2(g)} \rightarrow 2LiBH_{4(l)} + MgH_{2(s)} \quad (17)$$

$$2LiBH4_{(l)} + MgH_{2(s)} \rightarrow 2LiBH_{4(l)} + Mg_{(s)} + H_{2(g)} \quad (18)$$

$$2LiBH_{4(l)} + Mg_{(s)} \rightarrow 2LiH_{(s)} + MgB_{2(s)} + 3H_{2(g)} \quad (19)$$

For the hydrogenation process, temperature and pressures of about 350°C and 50 bar are required. These conditions for the formation of $LiBH_4$ are notably milder than that needed for the re-hydrogenation of $LiBH_4$ from LiH and B [79]. For the dehydrogenation process, temperatures above 350°C and hydrogen overpressures higher than 3 bar of hydrogen overpressure are required. These conditions are harsher than those predicted by the thermodynamics (**Table 2**). The hydrogen overpressure upon dehydrogenation assures the reversibility of the hydride system since MgB_2 is only formed under backpressure conditions. As seen in the reactions, $LiBH_4$ is in liquid state because this complex hydride undergoes from solid to liquid state at about 270°C [81]. Furthermore, mainly upon dehydrogenation the kinetic behavior is sluggish, particularly the second step described by reaction (19), taking more than 10 h for the full hydrogen release. Therefore, one of the most effective strategies was the addition of transition metal compounds [82]. Through this strategy, transition metal halides $TiCl_3$, TiF_4, NbF_5 among others, notably improved the dehydrogenation kinetic behavior of $2LiBH_4 + MgH_2$ hydride system. In this case, the transition metal halide additive interacts with $LiBH_4$ during the preparation stage during milling and subsequent heating to reach the operative temperature for hydrogen interactions. This interaction results in the formation of stable and nanostructure boride species such as TiB_2 and NbB_2 with similar hexagonal crystal structure as MgB_2. The *in situ* formed nanostructured transition metal boride species act as centers for the nucleation and growth of MgB_2, hence, accelerating the second step of the dehydrogenation, reaction (19).

Other example with the same $2LiBH_4 + MgH_2$ hydride system is the addition of TiO_2, leading to the *in situ* formation of core-shell $LixTiO_2$ nanoparticles. The mechanism of these core-shell nanoparticles is different from the one described for transition metal boride species. The core and shell of the nanoparticles are composed of $Li_{0.59}TiO_2$ and $Li_{0.59}TiO_2$, respectively. Upon hydrogenation and dehydrogenation, the *in situ* formed core-shell nanoparticles works as reversible Li^+ pumps, promoting the early decomposition of $LiBH_4$ and providing Li^+ for its rehydrogenation. Furthermore, the core-shell Li_xTiO_2 also improve the dehydrogenation of MgH_2, reaction (18). The addition of 1 wt.% of TiO_2 to $2LiBH_4 + MgH_2$ hydride system leads to hydrogen capacities of about 10 wt.% H_2, markedly shorter uptake (25 min) and release hydrogen times (50 min) and reduced activations energies at 400°C. The rate-limiting step for the hydrogenation process of the $2LiBH_4 + MgH_2$ doped with TiO_2 was one-dimensional interface-controlled mechanism (JMA, n = 1), while for the pristine material it was generally a diffusion controlled process [82]. As a novel approach, Puszkiel et al. interpreted the two steps dehydrogenation process by the combination of the JMA model with n = 1 for the fast MgH_2 decomposing, reaction (18), and the modified autocatalytic Prout-Tompkins model for the decomposition of $LiBH_4$/formation of MgB_2. Therefore, this autocatalytic process accelerated with the further formation of MgB_2. The nanostructured core-shell Li_xTiO_2 particles prompt the fast formation of MgB_2 seeds by acting as Li^+ sink/source for the early decomposition of $LiBH_4$ and the subsequent formation of LiH, respectively.

$Mg(NH_2)_2 + 2LiH$ was also investigated as a potential hydrogen storage material due to its hydrogen capacity of about 5 wt.% and operative temperature roughly 220°C [83]. This hydride system uptake/release hydrogen in two steps according to reactions (20) and (21):

$$2\,Mg(NH_2)_{2(s)} + 3LiH_{(s)} \rightarrow Li_2Mg_2NH_{3(s)} + LiNH_{2(s)} + 3H_{2(g)} \qquad (20)$$

$$Li_2Mg_2NH_{3(s)} + LiNH_{2(s)} + LiH_{(s)} \rightarrow 2Li_2Mg(NH_2)_{2(s)} + H_{2(g)} \qquad (21)$$

Several additives such as halides and hydrides were used to try to improve the kinetic behavior of this system [84–89]. However, it was found that the addition of just 0.1 mol of $LiBH_4$ improves not only the kinetic behavior, but also reduces the reaction enthalpy from 38.9 to 36.5 kJ/mol H_2 [31, 90]. Moreover, the 6 Mg $(NH_2)_2 + 9LiH + LiBH_4$ molar composition is the optimum one for the kinetic-destabilization effect. *In situ* formed complex amide-borohydrides like $Li_4(BH_4)(NH_2)_3$ and $Li_2(BH_4)(NH_2)$ are responsible for improving the kinetic behavior of the hydride system, according to reaction (22) [91]:

$$6\,Mg(NH_2)_2 + 9LiH + LiBH_4 \leftrightarrows 3Li_2Mg_2(NH)_3 + Li_4(BH_4)(NH_2)_3 + 9H_2 \quad (22)$$

Further improvement for the 6 $Mg(NH_2)_2$ + 9LiH + $LiBH_4$ was achieved by co-adding YCl_3 and Li_3N. The *in situ* formation of nanostructured YH_3 and YB_x upon milling and hydrogen interaction leads to faster kinetic behavior with reduced activation energy and capacities of about 4 wt.% at 90°C. On one hand, *in situ* formed $Li_4(NH_2)_3(BH_4)$ enhances the Li^+ transport, providing a source for the fast formation of the reacting species. On the other hand, nanostructured YB_x species contribute to the dissociation of H_2 [92].

Figure 11 depicts the concepts of the *in situ* catalyst formation applied to "destabilized hydride system." The complex hydride ABHy (A = metal, B = non-metal) or metal can react with a binary hydride, MHx (M = metal; different from A) to lower the thermodynamic stability by the exothermal formation of MB, thus $\Delta H_1 > \Delta H_2$. The *in situ* formation of catalytic species lowers the activation energy, i.e., $Ea_1 > Ea_2$, but does not alter the thermodynamic stability of the system.

4.4 Nanoconfinement

This strategy consists in confining the dimensions of hydride particles to sizes lower than 25 nm by introducing them into a nanoporous matrix. For a simple hydride formation/decomposition reaction (23), the contribution of the excess surface area given by the nanosize of the metal ($M_{(s)}$) and metal hydride ($MH_{2(s)}$) must be taken into account as part of the reaction enthalpy as described in reaction (24), where Vm is the molar volume, *r* particle radius and E(γ, r, Eads) is the surface energy term which depends on the surface free energies (γ) of the metal hydride and the metal particle, on the molar volumes of the two solid reaction partners, and on an additional energy term Eads, which takes into account that binding of H_2 at the surface of both the metal and the hydride reduce the respective surface energy by minimizing the excess of surface energy (γ) arising from not bound surface atoms. Therefore, the classical van't Hoff equation (Section 2.2, Eq. (4)) is corrected by the effects of nanoconfinement by replacing ΔH for ΔH', which takes into account the surface effects owing to the nanometric condition of the particles as shown in Eq. (25) [93].

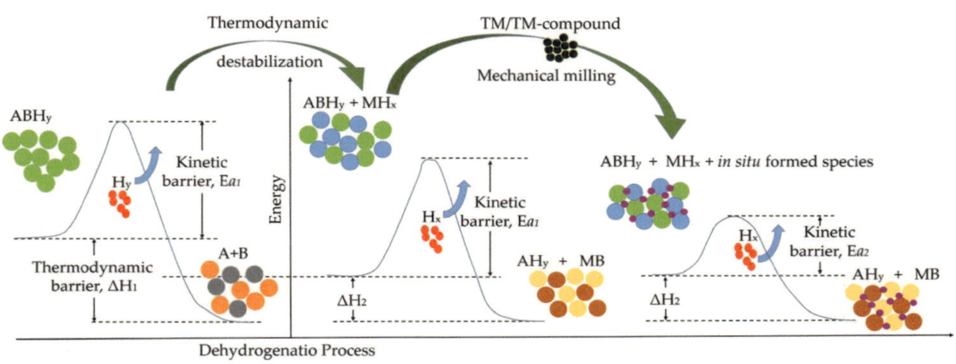

Figure 11.
Concept of in situ *catalyst formation applied to destabilized hydride system. Complex hydride = ABHy (A = metal, B = non-metal or metal); binary hydride, MHx (M = metal; different from A; TM = transition metal). ΔH = enthalpy,* Ea *= activation energy.*

$$M_{(s)} + H_{2(g)} \leftrightarrows MH_{2(s)} \tag{23}$$

$$\Delta H' = \Delta H + [(3Vm\ E(\gamma, r, Eads)/r] \tag{24}$$

$$lnP_{eq} = \left[\frac{\Delta H'}{RT}\right] - \left[\frac{\Delta S}{R}\right], \tag{25}$$

Nanoconfinement was basically proposed as a promising strategy to make hydride compounds reversible under mild conditions, as for example $LiBH_4$, $NaAlH_4$, and MgH_2, and to further destabilized hydrogen systems such as $2LiBH_4 + MgH_2$. However, in most of the cases, the main effect was observed on the kinetic behavior of the hydride compounds and systems. The nanometric range of the particles provide extremely large grain boundaries and notable shorter diffusion paths as well as optimized contacting for the materials, hence these properties account for improvements in terms of kinetic behavior and cycling stability. Carbon frames such as scaffolds, nanotubes, and aerogels are used for confining hydrides due to its light weight and inter condition. Despite the fact that this strategy usually leads to improved kinetic behavior without actually changing the thermodynamics of the hydride compound and system, the main constraint lay on the reduced capacity because of the introduction of the hydride into the nanoporous frame. Furthermore, it is hard to figure out in a possible scale-up for practical application [94–96].

5. Conclusions

Hydrogen can be considered as future energy carrier because of its large gravimetric energy density, abundance as combined with oxygen in water and its environmentally friendly condition at the time to release its energy. However, its low volumetric energy density owing to its gas conditions demands an efficient and a safe method to store hydrogen. One of the most attractive methods consists in storing hydrogen in solid state by forming a hydride compound. Much effort has been put in designing hydride forming materials and hydrogen storage systems based on hydride forming materials. One of the main constraints for the design of commercial solid-state hydrogen storage systems is the lack of a hydride forming material able to have proper thermodynamic stability as well as kinetic behavior. There are already materials with suitable experimental thermodynamic properties which would allow to reversibly uptake and release hydrogen at low temperature of about 90°C and under low pressures. Nonetheless, these conditions predicted by the thermodynamics of the system cannot be reached because of kinetic constraints. Therefore, strategies to tailor the kinetic behavior of hydride compounds and system of technical interest have been intensively developed. In this chapter, all the basics concepts about the chemical reaction, thermodynamics and kinetics for the right understanding of the solid-state hydrogen storage as hydride compounds or systems were explained. Then, the main strategies to tune the kinetic behavior of hydride compounds or systems were described. Among the most interesting strategies, milling process, transition metal, and transition metal compound doping, *in situ* catalyst formation and nanoconfinement were developed. Despite the fact that several hydrogen storage materials were notably improves by applying the exposed strategies, none of the materials nowadays fulfills the requirements for an efficient, a safe, and a cost effective practical application; hence, there is the need for further research in this field by applying the traditional design strategies and/or by developing new ones.

Acknowledgements

The author acknowledges CONICET (Consejo Nacional de Investigaciones Científicas y Técnicas) and Alexander von Humboldt Foundation (Fellowship Number: ARG-1187279-GF-P).

Conflict of interest

There is no conflict of interest.

Author details

Julián Atilio Puszkiel
National Council of Scientific and Technological Research (CONICET), San Carlos de Bariloche, Río Negro, Argentina

*Address all correspondence to: julianpuszkiel1979@gmail.com

References

[1] International Energy Agency. Oil Market Reports https://www.iea.org/media/omrreports/tables/2013-12-11.pdf, https://www.iea.org/media/omrreports/tables/2018-08-10.pdf

[2] www.worldenergy.org, https://www.worldenergy.org/wp-content/uploads/2016/10/World-Energy-Resources_FullReport_2016.pdf

[3] Schlapbach L, Züttel A. Hydrogen-storage materials for mobile applications. Nature. 2001;**414**:353-358. DOI: 10.1038/35104634

[4] Züttel A, Borgschule A, Schlapbach L. Hydrogen as a Future Energy Carrier. 1st ed. WILEY-VCH: Weinheim; 2008. 441 p. DOI: 10.1098/rsta.2010.0113

[5] Rand DAJ, Dell RM. Hydrogen Energy – Challenges and Prospects. 1st ed. Cambridge: The Royal Society of Chemistry; 2008. 296 p

[6] Schlapbach L, Züttel A. Hydrogen-storage materials for mobile applications. Nature. 2001;**414**:353-358. DOI: 10.1038/35104634

[7] Turner JA. Sustainable hydrogen production. Science. 2004;**305**:972-974. DOI: 10.1126/science.1103197

[8] Léon A. Hydrogen Technology: Mobile and Portable Applications (Green Energy and Technology). 1st ed. Berlin: Springer; 2008. 184 P. DOI: 10.1007/978-3-540-69925-5

[9] Graves C, Ebbesen SD, Mogensen M, Lackner KS. Sustainable hydrocarbon fuels by recycling CO_2 and H_2O with renewable or nuclear energy. Renewable and Sustainable Energy Reviews. 2011; **15**:1-23. DOI: 10.1016/j.rser.2010.07.014

[10] Becker WL, Braun RJ, Penev M, Melaina M. Production of FischereTropsch liquid fuels from high temperature solid oxide co-electrolysis units. Energy. 2002;**47**:99-115. DOI: 10.1016/j.energy.2012.08.047

[11] Ebbesen SD, Jensen SH, Hauch A, Mogensen MB. High temperature electrolysis in alkaline cells, solid proton conducting cells, and solid oxide cells. Chemical Reviews. 2014;**114**: 10697-10734. DOI: 10.1021/cr5000865

[12] Moller KT, Jensen TR, Akiba E, Li H-w. Hydrogen - A sustainable energy carrier. Progress in Natural Science. 2017;**27**(1):34-40. DOI: 10.1016/j.pnsc.2016.12.014

[13] Barthelemy H, Weber M, Barbier F. Hydrogen storage: Recent improvements and industrial perspectives. International Journal of Hydrogen Energy. 2017;**42**(11): 7254-7262. DOI: 10.1016/j.ijhydene.2016.03.178

[14] EU FP6 Integrated Project STORHY. http://www.storhy.net

[15] Sandrock G. A panoramic overview of hydrogen storage alloys from a gas reaction point of view. Journal of Alloys and Compounds. 1999;**293-295**:877-888. DOI: 10.1016/S0925-8388(99)00384-9

[16] S-i O, Nakamori Y, Eliseo JR, Züttel A, Jensen CM. Complex hydrides for hydrogen storage. Chemical Reviews. 2007;**107**:4111-4132. DOI: 10.1021/cr0501846

[17] Puszkiel J, Garroni S, Milanese C, Gennari F, Klassen T, Dornheim M, et al. Tetrahydroborates: Development and potential as hydrogen storage medium. Inorganics. 2017;**5**(4):74-97. DOI: 10.3390/inorganics5040074

[18] Puszkiel J, Andrade-Gamboa J, Gennari FC. Recent progress in Mg-Co-H and Mg-Fe-H systems for hydrogen energy storage applications.

In: Cheong KY, Impellizzeri G, Fraga M, editors. Emerging Materials for Energy Conversion and Storage. 1st ed. Cambridge: Elsevier; 2018. pp. 393-428. DOI: 10.1016/C2017-0-00444-5

[19] Garroni S, Santoru A, Cao H, Dornheim M, Klassen T, Milanese C, et al. Recent progress and new perspectives on metal amide and imide systems for solid-state hydrogen storage. Energies. 2018;**11**:1027-1055. DOI: 10.3390/en11051027

[20] Milanese C, Garroni S, Gennari F, Marini A, Klassen T, Dornheim M, et al. Solid state hydrogen storage in alanates and alanate-based compounds: A review. Metals. 2018;**8**:567-582. DOI: 10.3390/met8080567

[21] Walker G. Multicomponent hydrogen storage systems. In: Walker G, editor. Solid –State Hydrogen Storage – Materials and Chemistry. 1st ed. England: Woodhead Publishing Limieted; 2008. pp. 480-482. DOI: 10.1533/9781845694944.4.478

[22] Rudman PS, Sandrock GD. Metallurgy of rechargeable hydrides. Annual Review of Materials Science. 1982;**12**:271-294. DOI: 10.1146/annurev.ms.12.080182.001415

[23] Schwarz RB, Khachaturyan AG. Thermodynamics of open two-phase systems with coherent interfaces. Physical Review Letters. 1995;**74**(13): 2523-2526. DOI: 10.1103/PhysRevLett.74.2523

[24] Dantzer P, Meuner F. What materials to use in hydride chemical heat pumps? Materials Science Forum. 1988;**31**:1-18. DOI: 10.4028/www.scientific.net/MSF.31.1

[25] Stampfer JF, Holley CE, Suttle JF. The magnesium-hydrogen system. Journal of the American Chemical Society. 1960;**82–7**:3504-3508. DOI: 10.1021/ja01499a006

[26] Puszkiel J, Arneodo Larochette P, Gennari FC. Thermodynamic and kinetic studies of Mg-Fe-H after mechanical milling followed by sintering. Journal of Alloys and Compounds. 2008;**463**:134-142. DOI: 10.1016/j.jallcom.2007.08.085

[27] Mauron P, Buchter F, Friedrichs O, Remhof A, Bielmann M, Zwicky CN, et al. Stability and reversibility of $LiBH_4$. The Journal of Physical Chemistry. B. 2008; **112**:906-910. DOI: 10.1021/jp077572r

[28] Bogdanović B, Brand RA, Marjanović A, Schwickardi M, Tölle J. Metal-doped sodium aluminium hydrides as potential new hydrogen storage materials. Journal of Alloys and Compounds. 2000;**302**:36-58. DOI: 10.1016/S0925-8388(99)00663-5

[29] Chen P, Xiong Z, Luo J, Lin J, Tan KL. Interaction of hydrogen with metal nitrides and imides. Nature. 2002;**420**: 302-304. DOI: 10.1038/nature01210

[30] Vajo JJ, Skeith SK. Reversible storage of hydrogen in destabilized $LiBH_4$. The Journal of Physical Chemistry. B. 2005;**109**:3719-3722. DOI: 10.1021/jp040769o

[31] Xiong Z, Hu J, Wu G, Chen P, Luo W, Gross K, et al. Thermodynamic and kinetic investigations of the hydrogen storage in the Li-Mg-N-H system. Journal of Alloys and Compounds. 2005; **398**:235-239. DOI: 10.1016/j.jallcom.2005.02.010

[32] Christian JW. The Theory of Transformations in Metals and Alloys. 3rd ed. Amsterdam: Pergamon; 2002. 1170 p. In Press

[33] Khawam A, Flanagan DR. Solid-state kinetic models: Basics and mathematical fundamentals. The Journal of Physical Chemistry. B. 2006;**110**:17315-17328. DOI: 10.1021/jp062746a

[34] Stoch LS. Structural thermochemistry of solids. Themochim.

Acta. 1989;**148**:149-164. DOI: 10.1016/0040-6031(89)85211-6

[35] Gommel MM, Borkhart C, Fromm E. Absorption and desorption kinetics of hydrogen storage alloys. Journal of Alloys and Compounds. 1996;**238**:193-201. DOI: 10.1016/0925-8388(96)02217-7

[36] Brown ME. The Prout-Tompkins rate equation in solid-state kinetics. Thermochimica Acta. 1997;**300**:93-106. DOI: 10.1016/S0040-6031(96)03119-X

[37] Sharp JH, Brindley GW, Achar BN. Numerical data for some commonly used solid state reaction equations. Journal of the American Ceramic Society. 1966;**49**:379-382. DOI: 10.1111/j.1151-2916.1966.tb13289.x

[38] Jones LF, Dollimore D, Nicklim T. Comparison of experimental kinetic decomposition data with master data using a linear plot method. Thermochimica Acta. 1975;**13**:240-245. DOI: 10.1016/0040-6031(75)80085-2

[39] Höne G, Hemminger W, Flammersheim HJ. Differential Scanning Calorimetry. 1st ed. Berlin: Springer; 1996. 188 p. DOI: 10.1007/978-3-662-03302-9_3

[40] Kissinger HE. Reaction kinetics in differential thermal analysis. Analytical Chemistry. 1957;**29**:1702-1706. DOI: 10.1021/ac60131a045

[41] Wang XL, Suda S. A dehydriding kinetic study of $LaNi_{4.7}Al_{0.3}$ hydride by a step-wise method. Journal of the Less Common Metals. 1990;**159**:83-90. DOI: 10.1016/0022-5088(90)90135-7

[42] Yosephy Y, Ron M. Kinetic measurements of $LaNi_5H_x$ and $MnNi_{4.15}Fe_{0.85}H_x$ with constant pressure differential. Journal of the Less Common Metals. 1988;**147**:227-238. DOI: 10.1016/0022-5088(89)90196-3

[43] Reilly JJ, Yosephy Y, Johnson JR. Kinetics of the isothermal decomposition of lanthanum nickel hydride. Zeitschrift für Physikalische Chemie N.F. 1989;**164**:1241-1247. DOI: 10.1524/zpch.1989.164.Part_2.1241

[44] Goodell PD, Rudman PS. Hydriding and dehydriding rates of the $LaNi_5$-H system. Journal of the Less Common Metals. 1983;**89**:117-125. DOI: 10.1016/0022-5088(83)90255-2

[45] Bernauer O, Töpler J, Noreus D, Hempelman R, Richter D. Fundamentals and properties of some Ti/Mn based laves phase hydrides. International Journal of Hydrogen Energy. 1989;**14**:187-200. DOI: 10.1016/0360-3199(89)90053-0

[46] Han JS, Lee J-Y. A study on the dehydriding kinetics of Mg_2Ni intermetallic compound. Journal of the Less Common Metals. 1987;**128**:155-165. DOI: 10.1016/0022-5088(87)90201-3

[47] Kim SR, Lee JY. The effect of thermal cycling on the hydriding rate of $MmNi_{4.5}Al_{0.5}$. Journal of the Less Common Metals. 1990;**161**:37-47. DOI: 10.1016/0022-5088(90)90312-8

[48] Karlicek LF, Lowe IJ. Hydrogen diffusion in β-$LaNi_5$ hydride. Solid State Communications. 1979;**31**:163-165. DOI: 10.1016/0038-1098(79)90426-5

[49] Miyamoto M, Yamaji K, Nakata Y. Reaction kinetics of $LaNi_5$. Journal of the Less Common Metals. 1983;**89**:111-116. DOI: 10.1016/0022-5088(83)90254-0

[50] Song MY, Pezat M, Darret B. A kinetics study of the H-(Mg_2Ni-2.7wt.% Ni) system. Journal of the Less Common Metals. 1984;**103**:145-152. DOI: 10.1016/0022-5088(84)90373-4

[51] Fernández JF, Sánchez CR. Rate limiting step in the absorption and desorption of hydrogen by magnesium. Journal of Alloys and Compounds. 2002;**340**:189-198. DOI: 10.1016/S0925-8388(02)00120-2

[52] Jepsen J, Milanese C, Puszkiel J, Girella A, Schiavo B, Lozano GA, et al. Fundamental material properties of the $2LiBH_4$-MgH_2 reactive hydride composite for hydrogen storage: (II) Kinetic properties. Energies. 2018;**11**: 1170(15). DOI: 10.3390/en11051081

[53] Ushida HH, Fromm E: H_2 absorption rate of titanium films contaminated with increasing amounts of N_2, H_2O, CO, CO_2, CH_2 and SO_2. Zeitschrift für Physikalische Chemie N.F. 1989;**164**:1123–1128. DOI: 10.1524/zpch.1989.164.Part_2.1123

[54] Bershadsky E. Investigation of kinetics and structural changes in $TiFe_{0.8}Ni_{0.2}$ after prolonged cycling. Journal of the Less Common Metals. 1991;**172-174**:1036-1043. DOI: 10.1016/S0022-5088(06)80009-3

[55] Rudman PS. Hydrogen-diffusion-rate-limited hydriding and dehydriding kinetics. Journal of Applied Physics. 1979;**50**(11):7195-7199. DOI: 10.1063/1.325831

[56] Lozano GA, Ranong CN, Bellosta von Colbe JM, Bormann R, Fieg G, Hapke J, et al. Empirical kinetic model of sodium alanate reacting system (I). hydrogen absorption. International Journal of Hydrogen Energy. 2010;**35**: 6763-6772. DOI: 10.1016/j.ijhydene.2010.04.080

[57] Ley MB, Meggouh M, Moury R, Peinecke K, Felderhoff M. Development of hydrogen storage tank systems based on complex metal hydrides. Materials. 2015;**8**:5891-5921. DOI: 10.3390/ma8095280

[58] Suryanarayana C. Mechanical alloying and milling. Progress in Materials Science. 2001;**46**:1-184. DOI: 10.1016/S0079-6425(99)00010-9

[59] Huot J, Ravnsbæk DB, Zhang J, Cuevas F, Latroche M, Jensen TR. Mechanochemical synthesis of hydrogen storage materials. Progress in Materials Science. 2013;**58**:30-75. DOI: 10.1016/j.pmatsci.2012.07.001

[60] Abdellaoui M, Gaffet E. The physics of mechanical alloying in a planetary ball mill: Mathematical treatment. Acta Metallurgica et Materialia. 1995;**43**: 1087-1098. DOI: 10.1016/0956-7151(95)92625-7

[61] Puszkiel JA, Gennari FC, Arneodo Larochette P. Synthesis of $Mg_{15}Fe$ materials for hydrogen storage applying ball milling procedures. Journal of Alloys and Compounds. 2010;**495**: 655-658. DOI: 10.1016/j.jallcom.2009.10.011

[62] Puszkiel JA, Arneodo Larochette P, Gennari FC. Hydrogen storage properties of MgxFe (x: 2, 3 and 15) compounds produced by reactive ball milling. Journal of Power Sources. 2009; **186**:185-193. DOI: 10.1016/j.jpowsour.2008.09.101

[63] Zaluski L, Zaluska A, Ström-Olsen JO. Nanocrystalline metal hydrides. Journal of Alloys and Compounds. 1997; **253-254**:70-79. DOI: 10.1016/S0925-8388(96)02985-4

[64] Varin RA, Czuijko T, Wronsky ZS. Nanomaterials for Solid State hydrogen Storage. 1st ed. New York: Springer; 2009. . 325 p. DOI: 10.1007/978-0-387-77712-2

[65] Haraki T, Oishi K, Uchida H, Miyamoto Y, Abe M, Kokaji T, et al. Properties of hydrogen absorption by nano-structured FeTi alloys. International Journal of Materials Research. 2008;**99**:507-512. DOI: 10.3139/146.101669

[66] Puszkiel JA, Gennari FC, Arneodo Larochette P, Ramallo-López JM, Vainio U, Karimi F, et al. Effect of Fe additive on the hydrogenation-dehydrogenation properties of $2LiH + MgB_2/2LiBH_4 + MgH_2$ system. Journal of Power Sources.

2015;**284**:606-616. DOI: 10.1016/j.jpowsour.2015.02.153

[67] Bratanich TI, Solonin SM, Skorokhod VV. Mechanical activation of hydrogen sorption with intermetallic compounds $LaNi_5$ and TiFe in powder systems. International Journal of Hydrogen Energy. 1995;**20**(5):353-355. DOI: 10.1016/0360-3199(94)00062-5

[68] Chung HS, Lee JY. Hydriding and dehydriding reaction rate of FeTi intermetallic compound. International Journal of Hydrogen Energy. 1985;**10**(7–8):537-542. DOI: 10.1016/0360-3199(85)90084-9

[69] Kulshreshtha SK, Jayakumar OD, Bhatt KB. Hydriding characteristics of palladium and platinum alloyed FeTi. Journal of Materials Science. 1993; **28**(15):4229-4233. DOI: 10.1007/BF00351259

[70] Bogdanović B, Schwickardi M. Ti-doped alkali metal aluminum hydrides as potential novel reversible hydrogen storage materials. Journal of Alloys and Compounds. 1997;**253**:1-9. DOI: 10.1016/S0925-8388(96)03049-6

[71] Atakli ZÖK, Callini E, Kato S, Mauron P, Orimo S-I, Züttel A. The catalyzed hydrogen sorption mechanism in alkali alanates. Physical Chemistry Chemical Physics. 2015;**17**:20932-20940. DOI: 10.1039/c5cp01684c

[72] Barkhordarian G, Klassen T, Bormann R. Kinetic investigation of the effect of milling time on the hydrogen sorption reaction of magnesium catalyzed with different Nb_2O_5 contents. Journal of Alloys and Compounds. 2006;**407**:249-255. DOI: 10.1016/j.jallcom.2005.05.037

[73] Evard E, Gabis I, Yartys VA. Kinetics of hydrogen evolution from MgH_2: Experimental studies, mechanism and modelling. International Journal of Hydrogen Energy. 2010;**35**(17):9060-9069. DOI: 10.1016/j.ijhydene.2010.05.092

[74] Liang G, Huot J, Boily S, Van Neste A, Schulz R. Catalytic effect of transition metals on hydrogen sorption in nanocrystalline ball milled MgH_2–Tm (Tm=Ti, V, Mn, Fe and Ni) systems. Journal of Alloys and Compounds. 1999; **292**:247-252

[75] Barkhordarian G, Klassen T, Bormann R. Catalytic mechanism of transition-metal compounds on Mg hydrogen sorption reaction. The Journal of Physical Chemistry. B. 2006;**110**(22): 11020-11024. DOI: 10.1021/jp0541563

[76] Barkhordarian G, Klassen T, Bormann R. Fast hydrogen sorption kinetics of nanocrystalline Mg using Nb_2O_5 as catalyst. Scripta Materialia. 2003;**49**:213-217. DOI: 10.1016/S1359-6462(03)00259-8

[77] Friedrichs O, Aguey-Zinsou F, Ares Fernandez JR, Sanchez-Lopez JC, Justo A, Klassen T, et al. MgH_2 with Nb_2O_5 as additive for hydrogen storage: Chemical, structural and kinetic behaviour with heating. Acta Materialia. 2006;**54**:105-110. DOI: 10.1016/j.actamat.2005.08.024

[78] Friedrichs O, Martinez-Martinez D, Guilera G, Sanchez-Lopez JC, Fernandez A. In situ energy-dispersive XAS and XRD study of the superior hydrogen storage system MgH2/Nb2O5. Journal of Physical Chemistry C. 2007;**111**(28): 10700-10706. DOI: 10.1021/jp0675835

[79] Züttel A, Wenger P, Rentsch S, Sudan P, Mauron P, Emmenegger C. $LiBH_4$ a new hydrogen storage material. Journal of Power Sources. 2003;**118**:1-7. DOI: 10.1016/S0378-7753(03)00054-5

[80] Bösenberg U, Doppiu S, Mosegaard L, Barkhordarian G, Eigen N, Borgschulte A, et al. Hydrogen sorption properties of MgH2-LiBH4 composites. Acta Materialia. 2007;**55**:3951-3958. DOI: 10.1016/j.actamat.2007.03.010

[81] Nakagawa T, Ichikawa T, Hanada N, Kojima Y, Fujii H. Thermal analysis

on the Li-Mg-B-H systems. Journal of Alloys and Compounds. 2007;**446**: 306-309. DOI: 10.1016/j.jallcom. 2007.02.097

[82] Bösenberg U, Kim JW, Gosslar D, Eigen N, Jensen TR, Bellosta von Colbe JM, et al. Role of additives in $LiBH_4$-MgH_2 reactive hydride composites for sorption kinetics. Acta Materialia. 2010; **58**:3381-3389. DOI: 10.1016/j. actamat.2010.02.012

[83] Leng HY, Ichikawa T, Hino S, Hanada N, Isobe S. New metal-N-H system composed of $Mg(NH_2)_2$ and LiH for hydrogen storage. The Journal of Physical Chemistry. B. 2004;**108**: 8763-8765. DOI: 10.1021/jp048002j

[84] Torre F, Valentoni A, Milanese C, Pistidda C, Marini A. Kinetic improvement on the CaH_2-catalyzed Mg $(NH_2)_2$ + 2LiH system. 2014;**645**:S284-S287. DOI: 10.1016/j.jallcom.2014.12.228

[85] Amica G, Enzo S, Larochette PA, Gennari FC. Improvements in the hydrogen storage properties of the Mg $(NH_2)_2$ –LiH composite by KOH addition. Physical Chemistry Chemical Physics. 2018;**20**:15358-15367. DOI: 10.1039/c8cp02347f

[86] Liang C, Liu Y, Wei Z, Jiang Y, Gao M. Enhanced dehydrogenation/hydrogenation kinetics of the Mg $(NH_2)_2$-2LiH system with NaOH additive. International Journal of Hydrogen Energy. 2011;**36**:2137-2144. DOI: 10.1016/j.ijhydene.2010.11.068

[87] Li C, Liu Y, Gu Y, Gao M, Pan H. Improved hydrogen-storage thermodynamics and kinetics for an RbF-doped $Mg(NH_2)_2$-2 LiH system. Chemistry – An Asian Journal. 2013;**8**: 2136-2143. DOI: 10.1002/asia.201300323

[88] Li C, Liu Y, Ma R, Zhang X, Li Y. Superior dehydrogenation/hydrogenation kinetics and long-term cycling performance of K and Rb Cocatalyzed Mg $(NH_2)_2$-2LiH system. ACS Applied Materials & Interfaces. 2014;**6**: 17024-17033. DOI: 10.1021/am504592x

[89] Cao H, Wang H, He T, Wu G, Xiong Z, Qiu J, et al. Improved kinetics of the $Mg(NH_2)_2$–2LiH system by addition of lithium halides. RSC Advances. 2014;**4**: 32555-32561. DOI: 10.1039/C4RA02864C

[90] Hu J, Liu Y, Wu G, Xiong Z, Chua YS. Improvement of hydrogen storage properties of the Li–Mg–N–H system by addition of $LiBH_4$. Chemistry of Materials. 2008;**20**:4398-4402. DOI: 10.1021/cm800584x

[91] Wang H, Cao H, Wu G, He T, Chen P. The improved hydrogen storage performances of the multi-component composite: $2Mg(NH_2)_2$-3LiH-$LiBH_4$. Energies. 2015;**8**:6898-6909. DOI: 10.3390/en8076898

[92] Cao H, Zhang W, Pistidda C, Puszkiel J, Milanese C, Santoru A, et al. Kinetic alteration of the $6Mg(NH_2)_2$–9LiH–$LiBH_4$ system by co-adding YCl_3 and Li_3N. Physical Chemistry Chemical Physics. 2017;**19**:32105-32115. DOI: 10.1039/C7CP06826C

[93] Fichtner M. Properties of nanoscale metal hydrides. Nanotechnology. 2009; **20**:204009-204013. DOI: 10.1088/0957-4484/20/20/204009

[94] Vajo JJ. Influence of nano-confinement on the thermodynamics and dehydrogenation kinetics of metal hydrides. Current Opinion in Solid State & Materials Science. 2011;**15**:52-61. DOI: 10.1016/j.cossms.2010.11.001

[95] Nielsen KN, Besenbacher F, Jensen TR. Nanoconfined hydrides for energy storage. Nanoscale. 2011;**3**:2086-2098. DOI: 10.1039/C0NR00725K

[96] Yu X, Tang Z, Suna D, Ouyang L, Zhu M. Recent advances and remaining challenges of nanostructured materials for hydrogen storage applications. Progress in Materials Science. 2017;**88**: 1-48. DOI: 10.1016/j.pmatsci.2017.03.001